D1827128

International Patent Litigation

Developing an Effective Strategy

Consulting Editor **David Wilson**

Consulting editor
David Wilson

Publisher
Sian O'Neill

Marketing manager
Alan Mowat

Production
John Meikle, Russell Anderson

Publishing directors
Guy Davis, Tony Harriss, Mark Lamb

International Patent Litigation: Developing an Effective Strategy
is published by
Globe Law and Business
Globe Business Publishing Ltd
New Hibernia House
Winchester Walk
London Bridge
London SE1 9AG
United Kingdom
Tel +44 20 7234 0606
Fax +44 20 7234 0808
Web www.gbplawbooks.com

Printed by CPI Antony Rowe, Chippenham, Wiltshire

ISBN 978-1-905783-21-2

International Patent Litigation: Developing an Effective Strategy
© 2009 Globe Business Publishing Ltd

DISCLAIMER
This publication is intended as a general guide only. The information and opinions which it contains
are not intended to be a comprehensive study, nor to provide legal advice, and should not be treated
as a substitute for legal advice concerning particular situations. Legal advice should always be sought
before taking any action based on the information provided. The publishers bear no responsibility for
any errors or omissions contained herein.

Table of contents

Preface

David Wilson
Herbert Smith LLP

"It was the best of times, it was the worst of times,…"
Charles Dickens (*A Tale of Two Cities*)

As the importance of technology-based products and processes continues to increase, so too does the significance of patents. Patents are extremely powerful – they provide invaluable market exclusivity to their owners and present a potential barrier to third parties' freedom to operate. Unsurprisingly, therefore, patent-related disputes are becoming more common and their consequences more significant than ever. This increase in the need to effectively resolve patent related disputes has led most jurisdictions to develop specialised procedures and, in some instances, tribunals, to be able to deal with the unique nature of patent litigation. Although, much like other areas of law, patent law and practice have their own special features: perhaps unlike any other area of law, patent litigation involves understanding the subject matter of the dispute at a very detailed level, often equivalent to those who are experts in the relevant field. Moreover, the field is not limited to one particular area of science or technology. Those having to resolve patent disputes can find themselves having to understand the leading-edge technologies in almost all industry sectors, from heavy engineering to nanotechnology, from pharmaceuticals to food science. Another factor that forms part of the backdrop of how patent law develops is the recognition that a patent is essentially a bargain between an inventor and the state – in return for sharing the invention with society, the inventor is given an absolute monopoly over the invention, but for a finite period. The balancing act, between encouraging innovation through reward on one hand while not stifling competition on the other, is recognised throughout patent law, from the requirements of valid patent to the remedies granted for infringement, and with every balancing act there are various and varied ways in which the balance can be sought to be achieved.

As the chapters in this book will demonstrate, jurisdictions have developed their own ways of dealing with the issues that can arise in patent litigation. Having such specialised procedures in place can only be a good thing since they are designed to hopefully make litigating a patent easier, and hence cheaper. However, as will also be seen from the contents of this book, individual countries have developed their own individual methods of handling patent litigation and although some countries may choose to do things in a similar fashion in relation to some aspects, some of the differences are extremely significant (eg the bifurcation of infringement and validity proceedings compared with unitary systems). Hence while the development of specialist procedures for handling patent cases is to be applauded, the lack of any harmonisation in how these procedures have been developed is what makes litigating patents on an international scale very difficult if one is to maintain a consistent objective in enforcing one's patent rights or testing those of another.

This lack of consistency is hardly surprising – different jurisdictions have different systems for dealing with civil disputes, and the differences can be as fundamental as those between a common-law and a civil-law system. Moreover, patents are territorial in nature and unless and until we have a global prosecution system that grants to the applicant a 'global patent', the enforcement of patent rights will naturally tend to be looked at territorially.

However, that is not to say that there have been no efforts to harmonise international patent litigation, at least in respect of some parts of the world that already have a harmonised system of granting patents – although progress has been, to say the least, slow. Since 1977 it has been possible to apply for a European patent, as an alternative (or in addition) to a national patent application, by making a single application to the European Patent Office (EPO). This system, which involves a centralised prosecution system in which the applicant designates which member states it wishes its European patent to have effect in, has proved a great success. Most large companies that want to protect their inventions in Europe take the European patent route due to the cost effectiveness of only having to make a single application. However, the European patent system is based on the European Patent Convention (EPC) and although most of the 33 countries that are parties to the EPC are also members of the European Union, the EPC is not part of EU law and the European Patent Office is not an EU institution (one cannot appeal from the EPO to the European Court of Justice, for instance). Moreover, the EPC says very little about enforcement of European patents, this being a job for national courts, and while it does contain guidance as to how a European patent is to be construed for infringement purposes, the way in which this is implemented by the national courts can differ. Therefore, the situation of having a harmonised granting procedure in Europe, but no unified method of enforcing a granted patent, has led in recent years to the discussion of whether this can be remedied.

Views on how this can best be achieved are split. While the member states of the EPC have proposed the setting up of an integrated judicial system (known as the European Patent Litigation Agreement, or EPLA) that would have uniform rules of procedure for dealing with European patents granted by the EPO, together with a common appellate court, the European Commission has proposed the creation of a new patent right, the Community patent, together with a Community Patent Court. Unfortunately, discussions concerning both proposals are currently deadlocked and look to remain so for the foreseeable future. Although there is a great deal of support from the European patent judiciary for a harmonised system, some of the sticking points relate more to the political aspects of the system than to the practical requirements. It therefore remains to be seen whether compromise can be reached and, if it is, whether the level of compromise results in a system that gains the confidence of the users of the system, for unless a unified system meets the needs of the parties involved rather than the lawyers and politicians, it will not be used.

So, given it has not been possible for a part of the world with a well-established harmonized grant system in place to agree on a unified enforcement procedure, the chances of this happening on a wider scale must be remote to say the least. Hence the current system of individual patent proceedings being necessary in each jurisdiction of interest is likely to remain for a good while yet. However, while this may have not been so much of a problem in the past, given the global nature of today's business world – in particular in respect of those business sectors which are technology based – parties to patent litigation now increasingly find themselves having to develop a patent litigation strategy on a global scale in the context of a system in which one has to litigate in each country separately and differently. This is not a straightforward task. Hence the purpose of this book is to provide an outline guide of patent litigation in various countries such that the points of difference and in common between them can be taken into account when developing an effective strategy.

This book is not intended to be a text book of patent law and practice for experts in the subject – there are numerous excellent books already available that perform this task. The purpose of this book is to provide a general guide to patent litigation in various jurisdictions so as to highlight the sorts of issues that should be considered when developing a coherent strategy for patent enforcement or challenge on an international scale. It is aimed primarily at those who could be involved in, or responsible for, managing patent litigation, rather than practitioners in a specific jurisdiction. General counsel for a company often finds themselves having to manage (and to report to the board about) a multi-jurisdictional patent dispute, even though they might not have had any previous experience of patent litigation. Likewise, in-house patent counsel (or those in private practice), who may have extensive prior experience of patent litigation in their 'home' jurisdiction, may find themselves involved in litigation in jurisdictions which do things very differently. The further aim of this book is therefore to give a comparative guide to the most common jurisdictions of interest (and in putting this book together I have been only too well aware that there will be some jurisdictions that have been omitted, either for reasons of space or deadlines, that people would find of interest) so as to illustrate how each differs in respect of those aspects which most affect how a strategy is developed. It is also intended to highlight which jurisdictions have certain advantages that can be utilised by a party and how to avoid the common pitfalls. At the very least it is hoped that this book provides a guide to what sort of questions one should be asking of those ultimately handling the litigation in each country, so that a plan can be put in place that minimises the risk of one set of proceedings having an adverse effect on another and maximises the chances of the litigation successfully meeting the commercial objectives of the business, whatever they might be.

David Wilson joined Herbert Smith as a partner in 2007 from a leading city IP firm where he had been a partner since 2001.

David's practice has focused on complex patent litigation involving multi-jurisdictional disputes and work involving technical subject matter in the life sciences sector and chemical industries. In addition, he advises clients in relation to trademarks, regulatory and non-contentious (contractual, licensing and corporate) matters.

David has a first-class honours degree in chemistry and, prior to pursuing a career as a lawyer, he worked in the chemical and life sciences (immunodiagnostics) industry. David also qualified as a barrister before joining his previous firm in 1994. He is recognised by the major legal directories as a leader in the intellectual property and, in particular, patent litigation fields.

Developing a strategy and managing international patent litigation

David Wilson
Herbert Smith LLP

1. Introduction

As will be seen from reading the other chapters of this book, different countries approach patent litigation differently, both from a procedural point of view and substantively. As a result, not only are the procedures by which patent disputes are resolved different, but different remedies can be available and the length of time to obtain resolution can vary widely. Moreover, anyone who has been involved in major patent litigation will know that it is not a task to be undertaken lightly. It will usually involve a team comprising lawyers, patent attorneys, technical experts and possibly witnesses of fact. These team-members may be in-house employees of the party considering litigation, external advisers or a mixture of both. Often, multiple teams will be needed to handle proceedings in different jurisdictions in parallel. Therefore, on the assumption that one has a choice of jurisdictions in which one could seek to enforce patent rights, what sort of issues should one consider when trying to decide what to do? While recognising that each situation will be different, the purpose of this chapter is to provide a general guide to the points to consider before (and the word 'before' cannot be stressed enough) commencing patent litigation, both from a commercial point of view and in terms of practical issues that can arise. The hope is that by ensuring these points are considered when developing a litigation strategy, the commercial goals of the litigation are achieved in as cost-effective a manner as possible. As such, the purpose of this chapter is to highlight the sort of questions one should be asking rather than the answers – they will need to be found either by looking at the other chapters in this book or, if matter-specific, seeking appropriate advice.

Of course, one cannot consider how to go about enforcing one's patent rights unless one has some rights to enforce. How one effectively develops a portfolio of patent rights that best suits the needs of one's business is beyond the scope of this book, but since it clearly forms the foundation of one's enforcement strategy, and to a certain extent predetermines some of the issues discussed here, it is something that should be given careful consideration from the outset of any business. A business may have its own in-house team responsible for drafting and prosecuting its patent applications or it may use external specialists. In some jurisdictions these people may be the same as those responsible for conducting any patent litigation, while in others the two roles may be, to a greater or lesser extent, separate disciplines carried out by differently qualified professionals. Whatever the situation, those responsible for the securing of patent rights should form part of the litigation team, since not only may they have valuable experience of, or information relating to, the patent being considered, but issues that arise during the litigation may need to be taken into account in how the business' patent portfolio is managed in the future.

2. Developing an effective strategy

2.1 Define your commercial goals

For a patentee (as opposed to a patent lawyer), patent litigation is not something carried out because it is interesting and enjoyable. It is a means to an end, and it is vital that a patentee determines what that end is, as not only will this determine in which of the possible jurisdictions litigation should be carried out, but also the remedies that should be sought. For instance, if the potential infringement is occurring in a jurisdiction in which the patentee does not itself have a commercial presence, then a strategy directed towards the possibility of the sides negotiating a licence may be more sensible than one in which the focus is securing an interim injunction as soon as possible with a view to making it permanent. Likewise, if the infringement involves a product or process which the patentee is widely exploiting such that it forms a vital part of the business, focus may need to be directed on preventing the infringement as quickly as possible, and this may override the need to recover monetary damages. Further, an existing licensing programme, together with the interests of the licensee(s), may have an impact on the litigation strategy. For example, who is actually suffering the damage caused by the infringement? Does the fact that the patent has already been licensed (say, on an exclusive or sole basis) limit the potential for any negotiated settlement? If so, does this apply to all jurisdictions? In the case of a company accused of patent infringement, how important to the business is the accused product or process? Is it worth fighting the dispute all the way to trial? Can alterations be made to the alleged infringement which would avoid the issue? Would any settlement (undertakings, etc) carry implications for any other part of the business (eg is the technology accused of infringement used in any other products or processes of the business? Has the business given warranties or indemnities in relation to the alleged infringement?). Does the patentee have any other patents which may prove problematic in any event, such that any settlement should take these into account too? In the case of a company that has identified patents belonging to another which may impact on the company's intended activities, is it possible to pre-empt the issue by applying for a declaration of non-infringement or revocation of the patent? If so, what are the requirements, if any, which must be satisfied? Which jurisdiction(s) should be considered? Will the matter essentially become a patent infringement case? If so, what are the advantages to having provoked such a scenario? Does it send a message to the marketplace? Can the patent be challenged in the granting office (eg European Patent Office opposition proceedings or re-examination proceedings in the US Patent Office)? If so, will this have any effect on the court-based options? Is the patent being exploited by the patentee? Would the patentee be likely to consider granting a licence on economically acceptable terms (which do not include just the royalty rate, but also territory and the scope of the licence)?

Once the commercial aims of the company contemplating patent litigation have been determined, it is crucial that these be communicated to, and understood by, any external counsel. Remember that all too often lawyers can have a tendency to pursue every argument, no matter how strong or weak, and that it may be necessary to remind them of the purpose of the litigation should the litigation start to raise interesting legal issues or side-disputes. There is little to be gained by being a party in a jurisdiction's leading case on the requirements for pre-action disclosure unless the outcome of the issue was important to the party's ultimate aims in commencing the litigation.

2.2   Assess the landscape

Prior to approaching the issue of what, if any, action should be taken in which jurisdiction, the first obvious step is to determine which patents one has available to deploy. In the case of a company concerned about its freedom to operate in a certain sector, patent searches should be carried out (such searches can be expensive and throw up a lot of irrelevant results unless the parameters of the search are properly tailored to the concern in question – specialist search agencies can advise, together with counsel, on the appropriate and most cost-effective manner in which such searches can be carried out). Once the potentially relevant patents have been identified, a preliminary assessment of the relative strengths and weaknesses of each should be carried out. Remember that different jurisdictions have different attitudes to the same sort of patent (both based on the claimed subject matter and the way in which is it is claimed) and local advice may need to be sought. A balance may need to be struck between the commercial importance of a jurisdiction and its court's attitude to the sort of patent in question.

2.3   Where do you litigate?

The choice of which jurisdiction(s) in which to litigate is obviously one of the fundamental decisions that has to be made in any patent-related dispute. In today's commercial world most patentees hold equivalent patents in at least several key jurisdictions, and in any major case infringement is likely to be happening in more than one country. Discussed below are some of the points that should form part of the decision-making process and of course the issues discussed in each of the other chapters in this book may be relevant to a greater or lesser extent. Advice from appropriately qualified lawyers in some of the potentially relevant jurisdictions may be required in order to reach a final decision on this issue, but this itself raises an issue: in which countries do you choose to instruct local lawyers, given that the purpose of the advice is to decide which countries to choose? Clearly seeking such advice in all countries in which one holds patents would be prohibitively costly. However, external lawyers who are experienced in international patent litigation should be able to provide preliminary advice on the advantages and disadvantages of the different jurisdictions to enable the selection of local advice to be focused on the leading candidate countries.

Therefore, on the assumption that one has a choice of jurisdictions in which to take possible action, either as a patentee or as someone concerned about a patent belonging to another, the following is a general checklist of the points that should be considered.

(a) Availability of pre-action evidence gathering and/or disclosure

Some jurisdictions have extremely powerful procedures that can be used to obtain inspection of a party's documents or premises in order to determine whether infringement is occurring. Some of these do not even require that you have a patent covering the jurisdiction in question, just evidence to suggest that activities carried out within that country are leading to infringement of a patent elsewhere. If you suspect that one of your patents may be being infringed, but lack evidence to support such an allegation, then consideration should be given to whether any of these procedures are available. If such remedies are available, what are the requirements and do you meet them? Can the results of such court-sanctioned investigations be exported to other jurisdictions or are they restricted to use in local proceedings only? Consideration should also be given as to whether the procedure obliges a party to commence substantive proceedings for infringement in that jurisdiction within a short time period or risk the procedure being held null and void (with any possible consequences). Further, does a patentee applying for such an order have to give any sort of cross-undertaking in respect of any damage that the party being investigated may suffer as a result of the procedure? If so, what is your likely exposure?

(b) Availability of interim relief (eg preliminary injunctions)

If this is an important factor in achieving the commercial objectives of the litigation, then consideration should be given as to whether it is even available in the relevant jurisdictions in patent cases and what the courts' attitude to, and requirements for, it are. Some jurisdictions are prepared to grant such relief as a matter of course, while others have tended to take the view that, since the damages for patent infringement are open to assessment on the basis of a reasonable royalty, there is no question of irreparable harm being suffered pending trial. Some jurisdictions' views will depend greatly on the industry sector involved and the way in which the relevant market functions. Thought should be given as to what sort of evidence will be required to support an application for interim relief. Will the required evidence be available, and if so, will providing it have any effect on the business (eg does it contain confidential information about the party's business that may become public) or on the future conduct of the litigation?

Other factors to be considered include whether such relief is possible on a without-notice (*ex parte*) basis. What length of delay on the part of the patentee is permitted before such relief becomes unlikely, both on a without- and with-notice basis? How long does a court take to determine such an application – some jurisdictions will issue a decision on an application for interim relief in a number of days, while others can take many months (by which time the harm sought to be prevented may already have occurred). Again, the requirement for a party to give an undertaking or post a bond, so as to compensate the injuncted party for any damage suffered should it win at trial, should be examined. What is the likely exposure under such an undertaking? Does the potential downside outweigh the harm to be suffered if an injunction is not granted?

(c) ***Timescale of the proceedings***

The time it takes to conduct a patent action from start to first-instance judgment can vary widely from jurisdiction to jurisdiction. One also needs to factor in any appeal proceedings, and the time for these can again vary widely (and the slowest jurisdiction to first-instance judgment may not have the slowest appeal procedure). The timetable of an action not only needs to be considered from a basic management perspective, but also strategically. One needs to be aware of which country is likely to issue a decision first, what is the likely outcome of those proceedings and whether it is likely to have any effect on any other pending actions.

Consideration should also be given as to whether a jurisdiction offers an accelerated procedure and whether it is appropriate for the case in question. For example, in the United Kingdom there are several types of procedure available for patent actions which will result in a decision quicker than if the standard procedure were used. These include so-called 'speedy trials' (in effect, the matter is so urgent it must jump the queue and be heard as soon as possible), 'streamlined' actions (in which the action is of a lesser level of complexity such that reduced disclosure, no experiments, limited evidence and a reduced trial length is appropriate) and summary judgments (in which the case turns on a point which can be determined without the need for a full trial). Each one has its own requirements and possible consequences which may render the procedure suitable or not for the case in question.

Conversely, the possibility of an action being slowed down should also be considered, either as something to be avoided, or used as a strategic tool. A common example of this in Europe is a national court's attitude to whether actions should be stayed pending the outcome of opposition proceedings in the European Patent Office which, if successful, will result in the patent being revoked in all designated states. The attitude amongst European national courts varies widely on this issue and the factors that each one will take into account are also different.

(d) ***Disclosure (discovery)***

Will litigation in a jurisdiction involve the compulsory disclosure to the other side of any documents you have which may be relevant to the issues in the action? Many jurisdictions, in particular those that have a civil legal system (as opposed to one based on common law) have no concept of disclosure – a party need only disclose to the court and the other side documents which it chooses to disclose. Others, most notably the United States and the United Kingdom, have comparatively extensive disclosure requirements. The compulsory nature of disclosure means that all documents falling under the rules as disclosable must be shared with the other side (and potentially made public), whether they are helpful, unhelpful or confidential. This could be a disadvantage to a particular jurisdiction (do you have documents written which, in retrospect, could have been better phrased?), or an advantage (do you suspect that the other side may have documents they would rather not have to share?); but whatever the pros and cons of disclosure, one should be aware that it is also one of the factors responsible for litigation in certain jurisdictions being more expensive than others.

If disclosure is going to be involved, are their any limitations placed on the parties on the use they may make of documents exchanged during disclosure? For instance, in certain jurisdictions (eg the United Kingdom) a party may only use a document given over in disclosure for the purposes of the proceedings in which it was provided. This means it cannot be used in other proceedings in the United Kingdom, let alone in any proceedings in another jurisdiction, and the penalties for any misuse of disclosed documents can be severe. Questions should also be asked as to whether these restrictions are ever lifted and if so in what circumstances. For example, if the document is read out at a court hearing open to the public, is it then available for use in other proceedings? If so, is there any way to prevent this?

Thought should also be given as to whether any documents likely to be disclosed are likely to contain information confidential to the party disclosing it and whether steps can be taken to protect this confidentiality. Given that the parties to patent litigation will often be competitors in the marketplace, one may want to try to limit the personnel within the other party that have access to the sensitive documents to those that are absolutely necessary for the conduct of the litigation. If each party has similar concerns in this regard, it should be possible to agree terms relating to the dissemination of the documents within each party, although in the United Kingdom it is not uncommon for the court to be asked to resolve disputes in this area.

(e)   *Evidence*
Different jurisdictions favour different forms of evidence upon which to resolve patent disputes. Some prefer oral testimony while others prefer a more document-based approach. Some rely heavily on the evidence given by expert witnesses, while others do not. Those jurisdictions that do use expert evidence in patent cases are divided between those that permit the parties each to call their own expert witnesses (although such witnesses will be independent of the parties) and those that prefer to use a single, court-appointed, expert. Lawyers from each jurisdiction will of course sing the praises of their own system, and indeed there are advantages to each method, be it speed, costs or the likelihood of the decision being technically sound, but how each approach is best suited to the case in question, or not, will need to be discussed.

Thought should also be given to whether a witness (both experts and fact witnesses) will be subject to cross-examination. A potential expert witness may be very cogent and persuasive on paper, but if that same witness is unable clearly and consistently to express those same views when under the stress of hostile questioning from the other side's lawyer, they will be of limited value if this is part of a jurisdiction's procedure.

Another factor to which thought should be given is how a court will treat experimental evidence, for example if a party has conducted experiments specifically for the litigation (eg an analysis of a product to show infringement, or a repetition of an item of prior art to demonstrate lack of novelty). Certain jurisdictions have procedural rules that govern the use of such evidence which can influence how the experiments are designed and can greatly affect the costs of the proceedings. For example, in the United Kingdom a party upon which such evidence is served can request that the work be repeated in its presence so that it can observe the experiment being conducted. This means not only that the experiment should be designed to be reproducible, but that it will be necessary to perform a repeat usually in the presence of lawyers representing both parties. If the experiments are lengthy or complex, this will obviously increase costs.

(f)   *Law*

Although it may seem obvious, a careful analysis of each jurisdiction's substantive and procedural patent law should be carried out to determine whether there are any particular advantages or potential pitfalls applicable to the case in question. As a comparison of the other chapters in this book will demonstrate, each country's approach to both contains its own unique aspects, even those that are supposed to operate under a harmonised system of law. For example, one of the most fundamental aspects of any patent litigation is how the claims of the patent will be construed, and the approaches and methods used to arrive at a claim construction vary widely, with the result that a claim may be construed differently depending on where it is litigated.

(g)   *The court*

Some jurisdictions have specialist courts that deal with patent cases and hence have judges who are familiar not only with the legal issues that often arise in such actions, but who are also familiar, or at least comfortable, with cases which often involve discussion of highly technical and specialist subject matter. Other jurisdictions assign patent cases to general civil or commercial courts. Patent litigation in some jurisdictions, notably the United States, may involve the use of a jury. Therefore one should consider whether, all other things being equal, the subject matter of the intended action is of a level of complexity that means one would be better served by litigating in a specialist court. For example, is the case one in which one's initial view of the merits changes as one understands more about the technology and the various factors that may have been relevant to those working in the field at the relevant time? If so, is the tribunal going to need assistance in order to appreciate this? Such considerations can be more relevant when considering applying for interim relief, when the opportunity fully to explain what the case is about can be limited given the summary nature of such hearings – one does not want to apply for a preliminary injunction to a non-specialist judge and come away with a judgment refusing the application on the basis that, on the limited evidence the court has seen, it thinks that the patent is invalid.

Thought should also be given to whether the decisions of a court will have any precedent value in other jurisdictions. Will the courts of other jurisdictions have regard to the judgment and, although not bound to follow it, treat it as of persuasive authority? Factors which can influence this include whether the proceedings in question are heard by a specialist patent court or include a thorough examination of the evidence (cross-examination of experts and so on). If the judgments of one of the jurisdictions being considered are looked at approvingly by others, then this needs to be taken into account when planning the litigation strategy. What is the likely outcome of the litigation in this jurisdiction? Which other jurisdictions are likely to pay attention to the result? The answers to these questions can help determine when and how best to fit litigation in this jurisdiction into the overall timetable of the litigation.

(h)   Available remedies

Although compliance with TRIPS (the Agreement on Trade Related Aspects of Intellectual Property Rights) by most jurisdictions ensures that there are some minimum standards in place in terms of the remedies available for patent disputes, there is still scope for variation. One such example is the basis upon which damages for infringement are calculated, which can vary significantly. Hence the remedies that are available in each jurisdiction, and how they are enforced, should be factored into any strategy to ensure that the commercial objectives of the litigation are capable of being met.

Consideration should also be given to the issue of whether an appeal by the losing party will have any effect on the enforcement of remedies. If so, this could be for a significant period and again affect whether the litigation achieves the commercial objectives of the party within an acceptable timeframe.

(i)   Costs

As will be seen from the other chapters of this book, the cost of patent litigation varies very markedly from jurisdiction to jurisdiction. Further, the position varies as to whether the losing party in a case pays the legal costs of the successful party. Some jurisdictions provide for such recovery, some have each side pay its own costs whatever the outcome, and some have a system where the winning party's recovery is according to a statutory scale rather than actual expenditure. Further, the factors that affect where a party will come on such a scale also vary across jurisdictions and can be dependent on decisions made in the early stages of litigation. Even within the European Union, where the Enforcement Directive (2004/48/EC) has introduced the concept of recovery of 'reasonable costs' from an unsuccessful party across member states, the view of what constitutes 'reasonable' can be very different.

Clearly, the cost of any intended litigation strategy will need to be factored in. Is the likely cost of the strategy greater than the possible rewards? If so, is there still a good reason to continue (eg for long-term gains which may not be reflected in the actual awards of damages made by the courts or by provoking a settlement agreement that has added value to other areas of the business)? As discussed below, whatever the impact of cost has on the strategy, it is very unlikely that the party paying for the litigation (be it the actual party to the litigation or another part of the organisation) will write a blank cheque to the legal department – the cost of litigation will need to be budgeted for in the business and hence forecasts will need to be provided (including an assessment of how much of the expenditure may be recoverable from the other side if successful and when) at the outset and continually updated as the matter progresses.

2.4   Who should the parties be?

Once the possible jurisdictions in which to litigate have been decided, the next question, which can often be overlooked, is who should the parties be? If the holder of patent rights is considering an infringement action, thought should not only be given to who the defendant is going to be, but also who should be a claimant in the action. The obvious initial thought is that it needs to be the proprietor of the patent, and while this is usually correct, it may not automatically be the case. The patent may be the subject of an exclusive licence such that the licensee has the right to sue for infringement on its own. Some jurisdictions insist that even if this is the case, the patentee has to be made a party (either as a claimant or a defendant) to the action so as to prevent multiple actions relating to the same infringement. Further, if the patent is owned by a large international organisation, it is not uncommon for one part of the business to own the patent while it is another part (a separate but related corporate entity) that is responsible for exploitation of the invention. It is also not uncommon to find that there is no document recording the fact that the part of the business exploiting the invention is a licensee under the patent. This can raise its own issues – the patentee has the right to sue for infringement, but as it is not exploiting the invention, it is not suffering any harm and has no basis upon which to claim damages, etc. On the other hand, the part of the business (eg a local subsidiary) that is suffering harm in the market due to the infringement does not have the legal right to sue. This situation can be remedied by granting a formal licence to the subsidiary, but obviously this is something that should be checked and put in place if necessary before starting the litigation.

Even if the patentee can show that it is suffering harm due to the infringement (perhaps it receives income from its subsidiaries due to their exploitation of the invention which is being affected), it can sometimes be advantageous to include the relevant representative company in a jurisdiction as a claimant so as to give the action a local flavour for the court. This again may make a licence agreement necessary.

Care should also be given to those jurisdictions that require that any party purporting to have a right of action under a patent by way of a licence file a copy of that licence with the relevant patent office or court. These documents can be open to public inspection and the licence agreement may form part of a larger commercial arrangement which contains confidential information. While this can usually be dealt with by way of executing a short-form confirmatory licence agreement and filing it with the relevant authorities, quite often such documents must be notarised and other formalities complied with. This can take a surprising amount of time to complete, and hence it is sensible to consider whether any such arrangements could be necessary well in advance of the intended commencement of the litigation so that this does not turn out to be a frustrating source of delay.

In terms of who is to be a defendant, the law of the relevant jurisdiction will determine who it is that is carrying out an infringing act and hence who can be sued. However, if you are aware of who is selling the allegedly infringing product, it could still be worth investigating how the product comes to the market. Who is responsible for the manufacture or import of the product? Clearly if you can identify an act of infringement within the jurisdiction carried out by the person ultimately responsible for the product being available, then this is advantageous. It is also worth investigating the corporate structure of any companies that are infringing. Is a company merely a shell company against which a judgment in your favour is likely to be frustrated by the company's lack of assets? If so, is there another part of the organisation which can be joined as a defendant under principles relating to common design which allow you to target the part of the business who is ultimately responsible for the infringement? Also consider whether a defendant company is merely being used to provide a corporate shield for the activities of an individual (who could allow the company to simply collapse before continuing to infringe via another company). Does local law allow you to join the directors of such a company as personal defendants?

2.5 Appeals

An instinctive reaction to any unfavourable result is to appeal. However, the decision whether to appeal any such decision should be carefully considered. Is there an automatic right of appeal, or is permission from the lower court/appeal court required? How long will it take? What will it cost? Is the appeal a review of the first-instance decision, or is it a rehearing? Consider also whether the appeal has tactical value within the litigation strategy or potentially in settlement negotiations. Can the original decision, or rather its consequences, be stayed while the appeal is pending and if so, does this have any conditions attached?

3. Managing patent litigation – practical guidance

3.1 The initial stages

Often in-house counsel is a company's first port of call for initial advice as to whether there is a legal issue that may involve litigation. The commercial arm of the business may have identified a potential issue in the marketplace and ask for advice as to whether the company owns any patents that can be used to help resolve it. Obviously, the more in-house counsel is able to do in terms of identifying the relevant patents and forming a preliminary view on infringement and any potential weaknesses in the case, the easier it will be to instruct external lawyers. Depending on whether in-house counsel is a patent specialist or not, preliminary opinions on infringement and validity in each of the commercially important jurisdictions could be sought. Also, at this early stage, steps should be taken to identify any potentially relevant witnesses and to ensure relevant documents are preserved. If it seems possible that interim injunctions may be an option worth pursuing, it is absolutely essential to proceed without delay – in almost all jurisdictions in which preliminary injunctions are available in patent cases, the patentee must demonstrate that the matter is urgent. In this context any delay on the patentee's part in coming to court will need to be minimal and/or have a good explanation.

In the case of a potential defendant, often the first that in-house counsel will know of the issue is when a cease-and-desist letter, or perhaps even formal court process, arrives on the desk. It is absolutely essential that a team be put together and brought up to speed without delay. Formal court process often has to be responded to very quickly (eg within 14 days) and although this is little more than a formality in some cases, in others careful consideration has to be given to how these early stages are dealt with. Moreover, patent actions can now, in some jurisdictions, be brought to a full trial very quickly (in a matter of months rather than years) and there will be a lot of work to do from the start. Remember that the other side may have spent many months considering its options, putting its team together and preparing its case before starting proceedings. Because of this, it may also attempt to push forward the litigation as aggressively as possible. Hence, the need as a defendant to move fast.

3.2   Do you have enough time?

If you, as in-house counsel, are to be involved in a patent action, you will need to determine how much of your time is available to be devoted to it and how much of the action you wish personally to be responsible for. There are obviously many factors that affect this and some may be outside your control, but ideally the more time you can devote to the proceedings the better. It makes the external lawyers' jobs easier and hence helps reduce costs and, given in-house counsel will be closest to the commercial arm of the business, ensures that the action does not take on a life of its own and is focused on the objectives of the business.

If you are considering a major multi-jurisdictional piece of patent litigation, expect it to take up most, if not all, of your time. This may not be the case all of the time, but if periods of activity in several jurisdictions happen to coincide, which is not unusual, you will have to ensure not only that each case is properly managed, but that each case proceeds in a way that is consistent with the overall objectives of the litigation and that nothing done in one jurisdiction prejudices any of the other proceedings you may have ongoing or be considering. This is a full-time job and therefore it is quite often outsourced to external lawyers. It is not unusual to engage a firm specifically to act as a hub, managing the litigation in the various jurisdictions and reporting directly to in-house counsel. As discussed below, if this is done, it is essential to put in place an effective system of communication and for everybody involved to be clear as to who is responsible for decision making and providing instructions. Of course, this will increase costs, and so even if external counsel is to be engaged to manage multi-jurisdictional litigation, the more time in-house counsel can spend on the proceedings, the better.

It should also be appreciated that other employees of the patentee company may need to be ready to make themselves available to the litigation team as the matter progresses and often for significant periods. This is especially so for any inventors of the patent in question, whose role may range from acting as a technical adviser to being a testifying witness. Quite often the work that formed the basis of the patent will have been done years before the patent was granted, let alone litigated, such that the inventor will now be working on a different project or even have a different role within the company. This may mean that any involvement with the litigation is seen by the relevant person, or more usually, their superiors, as an interruption to their current role. It is therefore important for everyone involved to appreciate the importance of the litigation to the business and the need for that person to be able to spend time assisting in the litigation, without feeling as if doing so is adversely affecting their current position.

3.3 Putting the team together – selection of external lawyers

Choosing external counsel is very much like choosing someone to look after your car or carry out renovation work on your house – there is no substitute for personal experience or a recommendation from a trusted source. If one does not have any direct experience, it may be worth seeing whether you can talk to a contemporary in another company who has. Talk to as many people as possible. There are various guides and reviews of the legal market that now rank lawyers, both individually and by firm. If one has access to these, use them as a good starting point rather than making your ultimate choice based on these alone – they tend not to break down the lawyers by specific industry sector, and experience of the specific technology involved may prove valuable.

If it is intended that an external firm will take on the management of the international litigation and act as the main line of communication between the various "local" teams of lawyers and the client, then it is not necessary that litigation be proceeding in that country if there is another good reason to have that firm take on the management role of the project (such as experience in managing international patent litigation, pre-existing contacts in the various jurisdictions and so on). Likewise, it is not necessary to try to instruct one single firm that has a large number of international offices to conduct the entire case. Sometimes this can make communication and coordinating the matter easier, but not necessarily. Further, every office of the same firm may not have the same level of experience in patent matters. Moreover, the company may already have pre-existing relationships with lawyers in some of the jurisdictions being contemplated. Conflicts of interest with existing clients can also remove some firms from the equation. It is therefore suggested that the best solution is to review all possible candidates with a view to selecting the most suitable in each jurisdiction for the specific matter.

Once you have decided on a possible candidate or candidates, make sure you meet them and do not hesitate to ask any questions you have. Make sure you understand how the proposed team is made up, who is going to be doing what, and how personally involved in the matter will each team member be. Is the partner (who is usually the person who formed the basis of the recommendation) going to have sufficient time to devote to the matter? Obviously someone with as much experience as possible is preferred, but not if their practice is so busy that your matter will not receive the appropriate level of attention. Also, it is very important that you think you are going to be able to get on well with external counsel – you will be spending a lot of time with them and will want to develop a working relationship that allows you openly to ask questions of any aspect of the matter, including issues relating to fees or external counsel's conduct of the matter, without fearing that a question may be unpopular or inappropriate (no question should be either).

Make sure you know how much the matter is expected to cost. While it can be difficult to give exact forecasts for litigation, not least since it is dependent in part on the conduct of the other parties involved, experienced litigators should be able to provide you with an estimated budget for a matter. When comparing fees of different firms, especially hourly rates, not only should one remember that different jurisdictions are in effect separate markets when it comes to legal services (and so different factors drive what sets rates locally), but you should make sure you are comparing like with like. Different jurisdictions involve different amounts of work to get a patent action to trial, so looking at the specific chapters in this book should give you an idea of the average costs of patent litigation in each country. Within a jurisdiction, do not focus entirely on headline hourly rates – a firm that quotes lower headline rates for its fee earners may use a very large team of people on the matter such that its fees may end up larger than a firm that uses a smaller team but charges more per hour. Ask for a detailed estimated budget and makes sure it is clear how the figure has been arrived at – this will tell you how the lawyers see the action progressing and the various contingencies that have been considered and is likely to be more informative than just comparing hourly rates and total figures. Ask that the firm continue to review the budget and inform you as soon as it looks like it is no longer accurate, for whatever reason.

3.4 Communication and managing the team

As already mentioned, it is crucial to ensure that someone has responsibility to make it unambiguously clear what it is the various lawyers are doing, who is doing what, what the timetable is and who has ultimate responsibility. A carefully planned strategy must be executed in a consistent manner if it is to be effective, and hence ensuring that arguments put forward in one jurisdiction are, at the very least, not damaging to the position in other jurisdictions is vital. This may be the responsibility of in-house counsel, or an external lawyer may be given this role. If the litigation involves multiple cases in several jurisdictions it is critical that there is one person with overall responsibility who has sight of the entire picture, whether that be an external lawyer reporting to a client or in-house counsel. Quite often, although litigation may be being pursued in several jurisdictions, each one will have a different level of importance to the commercial objectives of the business. It is therefore important for each team to recognise that sometimes what is in the best interests of the litigation in one jurisdiction has to be forgone in the interests of the position in a jurisdiction of greater commercial or tactical significance. Hence the need for someone to have an objective view of the global litigation and how each jurisdiction fits into the picture.

Systems should be put in place to ensure that the relevant people in each jurisdiction are made aware of potentially relevant events taking place in other countries and/or cases that could have relevance to their own matter and likewise for them to be aware of anything in the proceedings in their jurisdiction that may be important or relevant to the proceedings elsewhere. Having a central hub to act as a clearing house of information can make this easier and avoid every member of the team reviewing everything (which will quickly cause the entire project to grind to a halt).

3.5 Disclosure issues

As noted above, certain jurisdictions, notably the United States and the United Kingdom, require the parties to litigation to disclose to the other side documents under their control which are relevant to the matters to be decided. This raises various issues, in particular where litigation in these jurisdictions is in parallel with others which have no concept of disclosure. The development of disclosure as part of the litigation process has also led to the development of various related concepts, such as privilege and how issues of confidentiality are dealt with. Jurisdictions which do not have disclosure as part of their litigation procedure may not have developed such concepts to such a degree or they may be different in their scope. Care should therefore be taken that actions in one jurisdiction do not prejudice the position being adopted in relation to disclosure in another. For instance, if in one jurisdiction disclosure of a whole class of documents is being resisted on the basis that those documents are privileged (and a class of documents may comprise many thousands of single documents), it is unlikely to be helpful if, in another jurisdiction, a document from this class is provided to the other party, or relied upon in court, unless it is made absolutely clear that this is without prejudice to the position being taken elsewhere. Even then the disclosure may amount to a waiver of rights elsewhere and so a coordinated approach must be taken at all times. Further, in those jurisdictions in which disclosure is the norm, systems can be put in place to limit the circulation of any documents which, although disclosable, contain confidential information. Agreements may be put in place between the parties or lawyers involved in the litigation, or they may be made the subject of a protective order of the court. Again, careless disclosure of any such documents in other jurisdictions may prejudice what can sometimes be the result of extensive negotiations as to terms of disclosure and who is to have access to the documents.

Also keep in mind that in US proceedings disclosure also extends to testimony by way of making potential witnesses available for deposition. Therefore, before deciding whether someone should be called as a witness, check to see whether they have already been deposed and if so, what the person said when questioned. If depositions have not yet been taken, make sure you are aware of when this is likely to happen and hence how it fits in with the timetables of parallel actions in jurisdictions in which that person may also be required to give evidence.

3.6 Trial

Trials, particularly in jurisdictions which employ an adversarial system and so emphasise the importance of witness testimony over documentary evidence, are extremely demanding. Until one has been through the process of the preparation and conduct of a major patent trial, few realise the amount of work that has to be carried out and the pressure that the entire team comes under. In those jurisdictions that have a specialist patent or technology court, and even in some that do not, the level of technical detail that is gone into, both by the lawyers and the court, can be surprising. The work to be managed includes the preparation of written and oral submissions; coordinating the attendance of witnesses and their management during trial (are they allowed to be in court when not testifying? Are they allowed to discuss the case with the team during breaks in their testimony?); ensuring that the witnesses are ready to give evidence (bearing in mind that the rules of professional practice that relate to what is permitted in terms of witness preparation vary widely from jurisdiction to jurisdiction); dealing with unexpected issues that come up during testimony; adjusting the legal analysis and submissions to reflect what emerges during the testimony and what can often be a changed perception of the facts as both sides' account becomes clearer; and dealing with interventions and questions from the court. In addition to all of this, settlement discussions may be taking place and there may also be enquiries from the press to be dealt with.

Given the above, it is important to make sure that someone is responsible for each task and is aware that it is their responsibility. Although external counsel may take on responsibility for most of the trial-related tasks, it is important that someone from the party involved is available throughout trial so that instructions can be given on issues as they arise. Consideration should also be made as to exactly when other members of the team should also be present – it can be useful to have the technical experts present while the technical witnesses called by the other side are giving evidence so as to assist in the preparation of cross-examination questions and closing submissions to be made in the light of the evidence given. If parallel proceedings are extant or being contemplated, copies of the draft written submissions should be circulated so as to ensure that the arguments are consistent with those that would need to be made in all jurisdictions. Likewise, local counsel should be warned in advance to be ready to have to review such submissions quickly, as often they are only finalised days before they are due to be filed. Consideration should be given as to whether it would be useful to have the lawyers from the other jurisdictions present to observe, or whether it is cost-effective simply to provide them with copies of the transcript of the hearing. With this in mind, care should be exercised in circulating documents or references to documents that are subject to confidentiality restrictions. Are all of the legal teams in other jurisdictions allowed to see them? Space should be made in people's diaries as soon as the trial date is fixed and educated guesses can be made as to who is needed when. Finally, it is unlikely that all of those that need to attend will be able to commute to the court from home, so responsibility for securing the necessary hotel accommodation should also be assigned. It is important that witnesses arrive in the witness box feeling refreshed and having had as good a night's sleep as possible.

3.7   Settlement

It will be an unusual case if the subject of possible settlement does not arise at some stage. While patent cases generally are arguably less prone to settlement than other commercial disputes (perhaps due to the nature of monopoly rights and the relief often sought), whether a patent case is a candidate for settlement will vary greatly depending on the commercial objectives of each side – is the argument simply about money (and if so, how far apart are the respective positions?) or is the issue one of the right to exclusivity? Further, different industry sectors often have different attitudes to the various possible avenues that exist for arriving at settlement (eg cross-licensing and so on) and these will often be mirrored in the commercial objectives a party has for bringing the litigation in the first place. Although those involved in the conduct of the patent litigation will of course need to have input on any settlement discussions, often discussions will be more productive if those ultimately responsible for them are not entrenched in the detail of the litigation and so can focus more readily on the commercial basis for any agreement. Of course they will need to be well briefed on the legal issues involved, the chances of success on each of them and their relative importance, so that those negotiating can determine the range within which an acceptable agreement lies.

During any consideration of possible settlement, and indeed for some industry sectors when considering the litigation strategy at the outset, it is crucial to consider whether the matter raises any competition law issues. The impact of monopoly rights such as patents, including how they are enforced and how they are exploited, is becoming of more and more interest to the authorities responsible for policing competition law, both nationally and internationally. In the patent field the opportunities to fall foul of competition law are legion and it can be very easy for those tasked with resolving a patent dispute to consider an agreement that looks like a commercially acceptable position for both sides, only for it to contravene antitrust requirements. This is a specialist area that requires specialist advice, and the temptation to think it is an issue that may never come up and so can be postponed until it does should be resisted. Often a competition-related concern can be taken into account in a way that still facilitates an acceptable settlement, while the possible consequences of ignoring the issue and operating under an arrangement which is later found to be anti-competitive can be very serious indeed.

3.8   Publicity

While one may have one's own view of whether any aspect of the litigation should be made public, it should be appreciated that this is only within the parties' control to a limited extent. While some jurisdictions do not publish to non-parties the fact that a case is pending or the decisions of cases, others do and recent decisions are read with interest by an active body of specialist press. It is therefore important to make sure that you have in place a strategy for dealing with publicity such that, in so far as is possible, you control the release of information to ensure that it is accurate – comments are often made by third parties on the basis of the limited information they have which turn out to be inaccurate once the full picture emerges.

If comment is to be made on a decision, or perhaps more importantly, on a pending piece of litigation, then careful thought needs to be given as to what can and cannot be said. Often, difficult questions of law will be involved in such situations. Further, specific comments about a case may also have an effect on the wider corporate objectives of the company and expectation management. It is therefore important that legal input to any public comment on the litigation takes place and that procedures are put in place to ensure that any pronouncements have been properly cleared by the necessary people. It should be remembered that inappropriate comment can at best be unhelpful to a case and at worst can prejudice its conduct and may even carry possible court-imposed sanctions such as contempt of court proceedings.

Consider preparing and securing approval for draft press releases before judgment is given in a case, based on the various possible outcomes. This will ensure that, as soon as a decision is made public, an accurate statement can be made as quickly as possible. Remember that in some jurisdictions you will receive little or no notice of the result of the litigation, and that the result may be price-sensitive.

Austria

Sascha D Salomonowitz
Schoenherr

1. Lawyers

1.1 Attorneys at law

Lawyers deal with both civil and criminal patent litigation in Austria. Since there is no distinction between solicitors and barristers, every qualified lawyer has a right of representation in patent matters. However, due to the complexity and special nature of patent cases, patent practice is dominated by practitioners specialising in intellectual property (IP).

The Austrian court system operates on a federal level (see Section 2 below) and all qualified lawyers (attorneys at law) may appear before any court throughout the country. Because only one court (the Vienna Commercial Court – *Handelsgericht Wien*) has jurisdiction to hear patent cases, all civil patent litigation is held in this forum, in Vienna. Likewise, all criminal patent litigation is handled by the Vienna Court for Criminal Matters (*Landesgericht für Strafsachen Wien*).

1.2 Patent attorneys

In contrast with other jurisdictions, lawyers and patent attorneys (ie patent agents) form completely separate professions. Patent attorneys are specially trained technicians who handle patent registration and prosecution (ie opposition and nullity proceedings). Rules of professional conduct forbid combined firms of lawyers/attorneys-at-law and patent attorneys.

1.3 Working together

In court cases, patent attorneys assist lawyers with regard to the technical aspects of the litigation and the patent, and this assistance can be invaluable. Civil litigation rules allow the patent attorney to be heard during the proceedings, but only the lawyer may plead or file submissions.

This kind of cooperation between lawyers and patent attorneys has the advantage for litigants of being able to cover both legal elements and specialist technical aspects of proceedings.

1.4 Summary

Patent infringement litigation in Austria is handled by lawyers, who are usually specialised IP practitioners. In most proceedings, patent attorneys will also participate on both sides and cover technical aspects of both the litigation and the relevant patent.

2. The court system

2.1 Exclusive jurisdiction of the Vienna Commercial Court

Civil patent matters in Austria are under the exclusive jurisdiction of the Vienna Commercial Court (*Handelsgericht Wien*). Currently, there are three departments within the Vienna Commercial Court which hear patent cases.

Cases are handled by panels consisting of three judges. One acts as presiding judge, and the others are a second judge (from the same court) and a so-called lay judge who is chosen from the list of lay judges. Usually, the lay judge in patent matters will be a patent attorney by profession and the presiding judge will try – where possible – to have a lay judge who is already a specialist in the relevant field. (Thus, a patent attorney known for handling pharmaceutical matters is likely to be chosen for a pharmaceutical patent case, or a patent attorney with a mechanical engineering background for a patent matter concerning an invention in the field of mechanical engineering.)

Moreover, since all patent matters are under the exclusive jurisdiction of the Vienna Commercial Court, the judges themselves are also very knowledgeable and experienced in the field of patent law.

2.2 Appellate courts

From the Vienna Commercial Court as court of first instance, any appeal is handled by the Higher Regional Court Vienna (*Oberlandesgericht Wien*) and the final instance is the Austrian Supreme Court (subject to certain procedural limitations).

2.3 Criminal courts

It should be noted that in Austria a wilful patent infringement is also a criminal offence, which is sanctioned by a so-called private criminal action in which the patent owner acts as *Privatankläger* analogous to the role of the state prosecutor. All criminal patent matters are under the exclusive jurisdiction of the Vienna Court for Criminal Matters (*Landesgericht für Strafsachen Wien*).

2.4 Foreign case law

Because comparatively few cases actually reach the courts (ie approximately 20 to 40 per year) and hence even fewer reach the Austrian Supreme Court, case law pertaining to patent law is rather limited. Whilst not bound by them in any way, litigants sometimes cite German decisions, as these have some persuasive effect due to certain similarities or known dissimilarities in the law.

2.5 Patent-related actions

(a) Declaratory action

Section 163 of the Austrian Patent Act allows for a declaration of non-infringement. Such actions are dealt with by the Austrian Patent Office. The purpose of these proceedings is to decide whether or not a certain object or process infringes a patent.

Hence, the Patent Office will decide whether a certain object or process would or would not infringe a patent if it was, *inter alia*, produced, marketed, offered or used (or respectively imported or owned with a view to being used). To this end, the Patent Office always carries out a comparison between the registered patent and the relevant object/process described in the request for declaratory judgment.

There is no obligation to stay infringement proceedings in the event of pending declaratory proceedings, and this has also been confirmed by the Supreme Court. However, there are certain arguments in favour of staying infringement proceedings in cases of pending declaratory proceedings. On the other hand, a request for declaration of non-infringement will be rejected if infringement proceedings are already pending at the time of service of the declaratory request (Section 163(3) of the Austrian Patent Act).

While the law allows both negative (declaration of non-infringement) and positive (declaration of infringement) requests, in practice the negative declaration is more important and more frequently used by a potential infringer. The patent owner will usually resort directly to a civil infringement action.

The only requirement for a declaration of non-infringement is that the applicant must already produce, use, import, etc the object/process, or have the firm intention to do so. The request must include a detailed description of the object/process, which will be attached to the decision.

A cease-and-desist letter which asserts a certain patent will usually suffice as a reason for such an action by the party addressed by such letter (ie the potential infringer).

However, in principle, this does not prevent a patent owner from asserting a patent before actually starting proceedings. Only an unjustified threat of patent infringement renders the threatening party liable for damages caused (and gives injunctive relief according to the Unfair Competition Act in cases of bad faith, as confirmed in recent case law of the Supreme Court: OGH 4 Ob 184/06x – Ophtalmoskop – ecolex 2007, 359). Not taking into account relevant prior art, wrong interpretation of the necessary inventive step, or wrong evaluation of the infringement may therefore lead to such damages. Case law is open on the question of whether actual knowledge of invalidity/non-infringement is necessary, or whether negligence suffices. Where an unjustified threat of patent infringement is directed against customers of the infringer, damages can also include damages for libel.

(b) Invalidity action

Invalidity actions are handled by the Austrian Patent Office. During civil infringement proceedings, the civil court may grant a stay for these separate invalidity proceedings by the patent office. Since the 2004 revision of the Austrian Patent Act, the court first has to determine the validity of the patent. According to this procedure, the court may also ask the Patent Office for a written opinion on validity based on the documents presented by the parties in the civil litigation. If the court sees a possibility that the patent may be found invalid, this will suffice to stay the civil proceedings.

Following the decision to stay the proceedings, the defendant must initiate a separate invalidity action at the Patent Office within one month. If such proceedings are not initiated, the court will continue the civil proceedings on the assumption that the patent is valid. Nevertheless, the court is still bound by a decision of the Patent Office before closing the first-instance proceedings. Even after a final judgment, a different decision on the question of validity is a valid reason for reopening the proceedings. Such motion must be filed within four weeks of the decision on validity.

2.6 Summary

The exclusively competent court for civil patent claims is the Vienna Commercial Court (*Handelsgericht Wien*). The competent authority for dealing with opposition and nullity proceedings is the Austrian Patent Office.

3. Procedure and timescale of proceedings

3.1 Civil infringement proceedings

Civil proceedings usually consist of a request for a preliminary injunction and then main proceedings. For first-instance decisions on requests for preliminary injunctions, a typical timeframe is between four and 12 weeks from filing.

It is possible to reach the Supreme Court even with a request for preliminary injunction, and a final decision (ie of the Appellate Court or the Supreme Court) on the preliminary injunction will usually be obtained in 'standard' IP matters within 10 months. Due to the higher complexity of patent matters and the fact that coordination of a three-person panel sometimes also adds a few weeks, the average time for a preliminary injunction in patent matters should be in the region of 12 to 18 months for a final decision. Inevitably, this will depend heavily on the complexity of the relevant matter.

In practice, many cases are settled after the final decision on the preliminary injunction. Although such decisions have no binding effect, they do have a certain persuasive power for the main proceedings.

If no settlement can be reached, the average time for the main proceedings in the first instance should be approximately one to two years, not including any time necessary for separate invalidity proceedings. (For cases where the invalidity of a patent is invoked and the court grants a stay for separate invalidity proceedings to be dealt with at the Patent Office, see Section 2 above.) Again, this timeframe will depend on the complexity of the matter and the time taken by court-appointed experts in providing their opinions.

An appeal in the main proceedings may be filed within a deadline of four weeks based on the following grounds:
- nullity of proceedings;
- violation of procedural law/due process;
- incorrect/incomplete determination of the facts and/or incorrect application/misinterpretation of the law.

It should be noted that Austrian civil procedure strictly limits the presentation of new facts on appeal. In appeal proceedings, no new facts can be asserted and no requests for taking additional evidence may be made. The appeal is a bilateral proceeding (ie opponents may file a reply to the appeal within a deadline of four weeks from receiving the appeal). Oral hearings on appeal are very rare (though possible) under civil procedure, and in most cases the Appellate Court will decide on the briefs of the parties. A final judgment (ie of the Appellate Court or the Supreme Court) will add one to two years to the timeframe.

There is no 'fast-track' approach in Austria. However, as stated earlier in this section, decisions regarding preliminary injunctions are quick and may help to resolve matters by settlement.

During preliminary injunction proceedings, both parties may use experts of their choice and will file an expert opinion on the question of validity, as well as on the issue of infringement, as permissible evidence to support the request for preliminary injunctive relief. Experts or other witnesses may not be formally interrogated during the preliminary injunction proceedings (see Section 7.1 below). During the main proceedings, the court will also appoint an expert to deliver an opinion on the question of validity and infringement for the main proceedings. Court-appointed experts can be declined in cases of conflict.

3.2   Declaratory actions

As described above (see Section 2.5), the Patent Act allows for a declaratory action, as regards both infringement and non-infringement. Such actions will usually take from two to four years for a final decision, depending on the complexity of the matter in hand.

3.3   Nullity proceedings

Nullity proceedings against a patent may be based on the same grounds as opposition proceedings, that is, lack of patentability, lack of disclosure of the invention (or, with regard to biological materials that have to be deposited in accordance with the Budapest Treaty on the International Recognition of the Deposit of Microorganisms for the Purposes of Patent Procedure, inaccessibility of the deposit) and inadmissible broadening (ic subject matter extending beyond the disclosure).

These proceedings consist of a written request, a counter-writ, pre-trial proceedings with possible further evidence gathering and finally an oral hearing, again with possible further evidence gathering. The decision of the board may declare the patent wholly or partially null. Due to the fact that the board will usually comprise experts in the relevant field, further expert testimony will not usually be requested by the panel.

3.4   EPO opposition proceedings

National nullity proceedings against the national part of a European patent will be stayed by the Patent Office itself whenever opposition proceedings on the same grounds are pending at the European Patent Office (EPO), and the national proceedings will be continued only if the EPO proceedings do not end with a decision on the merits.

3.5   Summary

Due to the highly specialised court and expert lay judges on the panel and the fact that the Supreme Court may be reached in the preliminary injunction phase of proceedings, decisions are usually comparatively fast and well researched. Nullity proceeding at the Austrian Patent Office usually take anywhere between two and four years and are set in a court-like procedure.

4. Availability of pre-action evidence gathering/disclosure

Following implementation of the EU Enforcement Directive, Austria's Patent Act (Sections 151a and 151b) includes provisions for orders for house searches, seizure of infringing goods and documents etc.

Since the entry into force of these provisions in mid-2006, there have been only a few applications, and no relevant case law of the Supreme Court has yet been published. Thus, it remains to be seen how practice develops for these orders.

Most importantly, the actual enforcement is carried out on a local court level by order of the Vienna Commercial Court, and the efficiency and speed of the evidence-gathering measures therefore also depend on the judge handling the matter at the local court.

From a procedural point of view, evidence gathering/disclosure efforts must be included in a request for a preliminary injunction, and these efforts need not necessarily be combined with proceedings on the merits at the outset. However, in order to fulfil the burden-of-proof threshold for (prima facie) validity and infringement, preparatory work is usually quite substantive and may well also be used immediately to file proceeding on the merits.

Results may be used freely and there is no apparent limitation to Austrian proceedings. However, the court is obliged to take into consideration the secrecy requirements of the defendant.

To a certain extent, criminal proceedings may also be used for information gathering. A wilful patent infringement is treated as a criminal offence by private criminal action, where the patent owner has the role of the state prosecutor. However, a recent far-reaching amendment of the Criminal Procedure Act (the actual title of which is *Strafprozeßordnung*) has somewhat changed and weakened the position of the private criminal action by requiring proof at the outset of the proceedings, and this is usually difficult to obtain. Practice continues to develop and it remains to be seen whether the private criminal action will return as a good and effective tool for information gathering.

4.1 Summary

Pre-action evidence gathering is available via a preliminary injunction according to the EU Enforcement Directive. Practice so far is limited and further developments remains to be seen. To a certain extent, criminal proceedings may also be used for information gathering.

5. Availability of interim relief (especially injunctions)

5.1 General availability

Interim relief is available in patent matters in the form of a preliminary injunction. Any infringement (or likely infringement) generally entitles a patent owner to a preliminary injunction. Austrian law does not have an urgency requirement for this request and preliminary injunctions may be requested even prior to the proceedings on the merits.

The applicant must show a well-founded case based on 'readily available evidence' (ie the court will not hear witnesses or expert testimony during the preliminary injunction proceedings). Applicants do not need to prove a danger of recurring acts, or to demonstrate irrevocable danger or loss.

If a preliminary injunction is requested without contemporaneously filing proceeding on the merits, the court will set a deadline for the filing of the proceeding on the merits. Due to the effort of preparing the request for a preliminary injunction, separate injunction proceedings and proceedings on the merits are rare.

5.2 Decisions without notice

Civil procedure and patent law allow applications to be decided without notice to the defendant. Previously, such requests have often been denied and have nevertheless been served upon defendants in the absence of extraordinary factors. With the implementation of the EU Enforcement Directive, the law now more clearly stipulates situations where a preliminary injunction must be issued without prior notice (in particular when prior notice will render the preliminary injunction useless, eg a house search or seizure order).

5.3 Permitted delay

There is no urgency requirement under the Austrian Civil Procedure Act/Patent Act/Execution of Judgment Act. Hence a preliminary injunction may be requested at any time, as long as injunctive relief is available for the patentee.

However, claims for damages (including rendering of accounts and information) are subject to a limitation period of three years from knowledge of the infringement and the infringing entity. While the claim for injunctive relief is not separately mentioned, the Supreme Court (OGH 15.5.1985, 4 Ob 317 – 318/95, ÖBl 1985, 129) has found that the statute of limitation of three years is also applicable to injunctive relief (and hence also to requests for preliminary injunctions). Of course, in the event of continued infringement a new deadline begins with each day of infringement.

For strategic purposes, it is easier to present an urgent case when the request for a preliminary injunction (and the claim as such) is filed expeditiously. It is also important to note the limitation period under the Unfair Competition Act for preliminary injunction requests based on slavish imitation. This period ends after six months of knowledge of the imitation and the imitator. It may be a worthwhile tactic also to base the claims on the general rules on slavish imitation, and hence to bear in mind the applicable statute of limitation.

5.4 Nullity proceedings

Preliminary injunction proceedings will not be stayed (Section 156(3) of the Patent Act). The court will decide whether the plaintiff has reasonably shown that the patent is valid and has been infringed (or whether the defendant has succeeded in countering) and will either grant the request for a preliminary injunction, or dismiss it.

5.5 Protective briefs

Austrian civil procedure does not recognise protective briefs.

5.6 Pros and cons of injunctions

A major argument in favour of injunctions is that they provide a comparatively quick decision by the same court which will later also deal with the proceedings on the merits. Appeal is also possible (subject to certain limitations) up to the Supreme Court and within a comparatively short timeframe.

Against this, the claimant must bear in mind the strict liability which applies where a preliminary injunction is unfounded. Under Section 394 of the Execution of Judgement Act (where the preliminary injunction is regulated), a patent holder is strictly liable in cases where it is later established that the injunction was awarded without reason (eg due to an invalid patent or non-infringement).

5.7 Summary

Interim measures are generally available in patent matters and are a speedy means of protecting a patent holder's rights.

6. Disclosure

Other than via measures of interim relief and/or private criminal action, Austrian civil procedure does not allow disclosure rules.

7. Evidence

7.1 Types of evidence

Under general Austrian civil procedure, evidence can take the following forms: witnesses, documents, expert witnesses, on-site inspections and parties' testimony. During the preliminary injunction proceedings, evidence needs to be readily available (ie in written form available to the court). In practice, it is possible to present 'informed persons' during preliminary injunction proceedings, and these are called in a relatively informal procedure to testify before court and will be examined only by the judge (usually even without lawyers present).

In proceedings on the merits, all forms of evidence are available: documents, written arguments, witnesses and expert opinions as well as expert witnesses giving oral testimony. During proceedings on the merits, witnesses must be heard in person by the court and written witness statements are not permissible.

7.2 Expert witnesses

Due to the technical nature of patent cases, expert witnesses play an important role in patent infringement proceedings. As described in Section 2 above, the panel of judges will usually already include a patent attorney. Moreover, in nearly all cases, there will be a court-appointed expert – usually another patent attorney – to provide an opinion on validity and infringement to the court.

Additionally, the parties will each file one or possibly several expert opinions (again written by patent attorneys) as to issues of validity and infringement.

7.3   Cross examination

Any witness (including expert witnesses rendering expert opinions) is regarded as evidence for either party. Either party may also interrogate each witness. Civil procedure does not formally distinguish between interrogation and cross-examination. In practice, the panel of judges will first interrogate each witness and then the lawyers will pose additional questions.

7.4   Experiments

Experiments may be conducted by each party and the results may be shown to the court – usually in the form of an expert opinion. Where a court-appointed expert also needs to conduct experiments, the court may ask the parties to provide the expert with the necessary material.

7.5   Summary

Austrian courts allow various types of direct evidence (witnesses, documents, expert witnesses, on-site inspections and parties' testimonies). Written witness statements are only permissible in preliminary injunction proceedings. Patent cases are characterised by extensive use of expert opinions on issues of validity and infringement.

8.   Law

8.1   Patent infringement

A patent infringement action can be undertaken in relation to every kind of (direct or indirect) patent infringement, that is, making, using, selling, offering for sale, or importing the patented invention (or an equivalent invention covered by the scope of the granted patent according to Section 22a of the Austrian Patent Act), without the authorisation of the patent owner. Claims construction is based on Article 69 of the EPC (Section 22a of the Austrian Patent Act).

8.2   Patent infringement in pharmaceutical patent matters

The mere obtaining of a market authorisation will usually not be regarded as a patent infringement in Austria, as long as the respective product is not actually produced or stocked for delivery in Austria (*Gamerith*, Der vorbeugende Unterlassungsanspruch, ÖBl 2005, 52). A 'timely' application for market authorisation should not be regarded as a patent infringement, due to the fact that the timeframe for obtaining market authorisation is not intended to extend patent protection. It will be up to the courts to interpret this question, but it may be argued that a very premature (and, hence, not timely) application for market authorisation could be regarded as a preparatory act for a patent infringement.

It is also important to mention in this context that the application for inclusion of a certain pharmaceutical product to the list of medications where the costs are borne by social security ("*Erstattungskodex*", red/yellow/green box system), where one of the preconditions for the listing is the ability to deliver the pharmaceutical, is a patent infringement, since this is considered as an offering of an infringing product (even if not a single pharmaceutical is actually sold). In practice, this is very important since pharmaceutical products can be widely marketed only if included in this list.

8.3 Bolar provision

The experimental use exception (Section 22(1) of the Austrian Patent Act) was introduced by the Patent Act Revision Act of 2005 (the so-called Bolar Amendment, which came info force on November 19 2005):

> "§ 22(1) The patent grants the patent owner the right to exclude others from producing, marketing, offering or using or importing or owning the subject matter of the invention for such purposes. The rights conferred by a patent do not extend to studies or trials as well as the consequential practical requirements, insofar as they are necessary for obtaining an authorization, approval or registration to market under pharmaceutical law..."

So far, there is no case law available from the Austrian Supreme Court regarding this provision and the (summarised) legislative history offers only limited guidance on the question. It explains that the introduction of the Bolar provision implements Article 10(6) of Directive 2001/83/EG and Article 13(6) of Directive 2001/82/EG (both as amended) and that the introduction of the exemption benefits Austria by allowing research on generics. Otherwise, the research would be done in other countries offering such exemption, or where the product is not patented (or patent protection has expired). The innovative pharmaceutical sector should therefore suffer no disadvantage from the introduction of this provision.

The main question is how the judges will define the requirement as to the necessity of the given studies/trial. According to the wording, it seems possible that a distinction might be drawn between early-stage and late-stage studies and trials, thereby interpreting the exemption to cover only those studies and trials which are directly necessary (ie arguably only late stage) for market authorisation under pharmaceutical law. In the absence of case law, only a court decision will finally show how the experimental use exception will be interpreted.

8.4 Assessment of patentability

Patents protect inventions in the field of technology and are protected through registration in the Austrian Patent Register. Patent applications must be filed in writing with the Austrian Patent Office. The legal requirements for registration are novelty of invention, an inventive step and susceptibility of industrial application.

According to Section 1(2) of the Austrian Patent Act, patents are not granted for certain specified inventions. Thus, the following are not considered to be inventions according to patent law: discoveries, scientific theories, mathematical methods, the human body, parts of the human body or the discovery of a part of the human body (including genetic sequences), designs, as well as concepts, rules and procedures for intellectual work, game concepts or computer programs and the mere reproduction of information.

Patents are also not granted for inventions the commercial exploitation of which would be contrary to public policy or morality, for example: methods for cloning human beings, methods for the modification of the human genetic code, the use of human embryos, or the creation and exploitation of hybrid creatures formed from a combination of human and animal cell material.

Moreover, patents are not granted for surgical or therapeutic techniques and methods for diagnosis with respect to the human or animal body; botanical genera and species (according to Article 5, Regulation 2100/94/EC on Community plant variety rights); breeds of animals; or biological (non-genetic) methods for breeding of plants or animals (in particular cross-breeding or selective breeding).

8.5 Summary

The Austrian Patent Act is a modern law which has implemented important EU directives, such as the Enforcement Directive.

9. Available remedies

9.1 Remedies

A patent gives the patent owner the right to exclude third parties from manufacturing, placing on the market, offering or using any product which is the subject of the patent. Furthermore, the importation or ownership of any product for such purposes is forbidden (Section 22 of the Patent Act).

Any production of infringing products in Austria is therefore treated as an infringement in Austria. The various methods of using a patent are generally treated equally by the Patent Act (also as regards the question of damages).

9.2 Financial remedies

Under Section 150 of the Patent Act, a patent owner is entitled to claim payment for the infringement of relevant IP rights. The patent owner has the right to reasonable compensation for the infringement (usually analogous to the grant of a licence). In cases of culpable patent infringement (ie knowingly using the patent (and there is no general obligation to check for patent infringements other than the usual good conduct of tradespeople in their trade)), the patent owner may instead also claim damages including lost profits or account of profits made by the infringer. Independent of any proof of damages, the patent owner may claim double damages (ie twice reasonable compensation for the infringement) in cases of gross negligence or intentional patent infringement.

9.3 EU Enforcement Directive

The EU Enforcement Directive has been implemented in Austria generally for IP matters and has considerably broadened the scope of pre-trial evidence gathering via house searches, seizure orders and so on. See Sections 4 and 5 above.

9.4 Effect of an appeal on enforcement

Whilst a preliminary injunction becomes immediately enforceable (in the absence of a contrary decision of the court, including certain possibilities for a defendant to post a bond/security payment), the decision on the merits usually only becomes enforceable once a final decision is rendered. However, the plaintiff may request a preliminary injunction at any stage of the proceedings on the merits.

9.5 Summary

Available remedies include:

- preliminary and permanent injunctive relief;
- removal and destruction of infringing materials;
- provision of information about the origin of infringing materials;
- publication of the judgment;
- rendering of accounts and payment of compensation or damages;
- criminal sanctions in case of intentional infringement.

10. Costs

10.1 General

The court filing fee depends on the amount in dispute. (This is rarely the actual amount claimed, but rather a "nominal" sum chosen by plaintiff.) Due to the fact that lawyers' tariffs recommend a minimum sum in dispute of €36,000, the court filing fee starts at €607 and may rise to 1.2%, plus €1,661, for sums above €363,360.

Further legal costs are attorneys' fees, which are calculated according to lawyers' tariffs (by fixed rates or hourly rates, depending on the fee arrangement). The work needed to conduct proceedings will usually depend on the complexity of the actual case resulting from the defence mounted by the defendant. Moreover, the number of court hearings, meetings and court briefs will also have an impact on lawyers' fees. Additionally, parties have to bear the costs of their expert witnesses and expert opinions, as well as the services of their patent attorney during the proceedings. Again, these will depend on the complexity of the technology and the number of technical documents to be reviewed.

10.2 Reimbursement

At the end of the proceedings the court will order the parties to reimburse the costs, based on the success of the claim. Such reimbursement is calculated on a statutory scale rather than being based on actual expenditure. However, external costs relating to parties' expert opinions and patent attorneys' services may be claimed. Where a claim is completely successful, the overall reimbursement of legal fees will usually be around 40% to 80% of actual costs accrued.

Similarly, the losing party will be liable to reimburse the opponent's costs using a statutory scale, rather than actual expenditure incurred.

10.3 Summary

The cost of patent litigation in Austria is comparatively low relative to other EU jurisdictions and the United States, with reimbursement of statutory costs for the successful party.

11. Hot topic(s)

11.1 Return to national litigation

After the *GAT v LuK* and *Roche v Primus* decisions of the European Court of Justice, the focus in patent litigation has returned to national litigation, as opposed to EU-wide injunctions which were previously granted in some jurisdictions (eg the Netherlands).

11.2 Reform of the Austrian Patent Act

The latest reform of the Patent Act has implemented a wide array of selective provisions, *inter alia*: changes relating to Regulation 2006/1901 on medicinal products for paediatric use and amending Regulation (EEC) No 1768/92 concerning the creation of a supplementary protection certificate for medicinal products; Directive 2001/20/EC on the conduct of clinical trials; Directive 2001/83/EC on the Community code relating to medicinal products for human use and Regulation (EC) No 726/2004 on the authorisation and supervision of medicinal products and establishing a European Medicines Agency; and recent revisions to the European Patent Convention.

11.3 Patent trolls/non-practising entities

With Germany playing host to the first lawsuit of a non-practising entity enforcing patents (the *IPcom* case), it is generally believed that a new trend will lead to more patent enforcement in the European Union and hence also in Austria. In the past, companies have been reluctant to pursue and enforce their patent portfolios, and an awareness of the value of unused patents is now beginning to influence corporate decisions.

Auctions of patents and other IP assets, as well as the use of intellectual property as collateral for corporate financing, will also lead to the creation of an IP marketplace, in which intellectual property receives an attributed financial value. The development of such a marketplace will allow easier appraisal of the value of patents and IP rights in general.

Austria summary

- Austrian courts offer a convenient forum for patent litigation, through specialised panels of judges.
- Preliminary injunction proceedings mean that court decisions can be obtained comparatively quickly, with the possibility of lodging an appeal up to the Austrian Supreme Court.
- Austria is part of all the major European and international treaties and has implemented relevant EU directives in the field of patent law.
- In addition to the measures implemented by the EU Enforcement Directive, Austrian criminal law offers the possibility of obtaining evidence by way of a private criminal action.

Belgium

Ignace Vernimme
Stibbe

---

1. Lawyers

The team in Belgian patent litigation consists of attorneys at law who may be assisted by one or more patent attorneys.

Attorneys at law are entitled to handle and plead the case in front of any court in which patent litigation occurs. Since attorneys at law are qualified in all Belgian jurisdictions, the intervention of a local lawyer is not required.

The patent attorney cannot plead in court and his intervention is, in principle, limited to drafting reports on the validity of the patent and on infringements. Exceptionally, patent attorneys are heard by the court. In such cases, the patent attorney does not act as witness, but as a representative of the party.

2. The court system

At first instance, five commercial courts (of Antwerp, Brussels, Ghent, Liège and Mons) are now exclusively competent to deal with patent litigation. This is with a view to improving the quality of judgments through increased specialisation. Thus, the civil courts, which traditionally deal with heterogeneous issues from divorce to intellectual property (IP) enforcement, will no longer be involved with patents. The fact remains, however, that the judges in the commercial courts are not specialised, do not in principle have a technical background and sometimes tend to favour a commercial rather than a strict IP approach.

An appeal can be launched before the Court of Appeal of the five aforementioned cities within one month of the formal notification of the decision. Unlike other jurisdictions, the Belgian judicial system does not provide the mechanism for leave to appeal. The judges in the Court of Appeal do not specialise in patent litigation, although there is a tendency to assign IP cases to specific chambers within the Court of Appeal.

Appeals against judgments of the Court of Appeal can be launched before the Belgian Supreme Court (*Court de Cassation*), which only rules on points of law and no longer on the facts of a case. In principle, it is not possible to leapfrog the Court of Appeal and to appeal the decision of the commercial court directly to the Belgian Supreme Court.

At first instance, the commercial court normally comprises one judge and the parties can ask for three judges in the appeal procedure.

There is a discussion in Belgian doctrine and jurisprudence regarding the extent to which Belgian courts should take note of decisions from other jurisdictions. Looking at the recent case law of several appeal courts, it is evident that the majority of the courts do not take into consideration the decisions of others jurisdictions. The courts of Antwerp, however, unlike the courts of Brussels and Liège, do take note of decisions from other jurisdictions. In practice, attorneys often submit decisions from other jurisdictions in an attempt to convince the judge to take them into account as one of the elements in his evaluation of the case.

The following patent-related actions are possible.

2.1 Descriptive seizure

This action allows a patent holder to obtain – in principle on a unilateral basis – from the president of the commercial court an order which allows the appointment of a judicial expert, who will be empowered to enter the business premises of the alleged infringer in order to prepare a description of apparatus, processes, products and documentation that are "likely to establish the alleged infringement". In principle, the judicial expert cannot provide in his report any advice as to validity of the patent or infringement. Within one month of the submission of the expert's report, the rights owner must launch a procedure on the merits relating to the alleged infringement. The president of the commercial court will grant an order on the basis of a prima facie valid IP title together with proof of a real risk of infringement. Penalties are generally included in the order in the event of non-compliance, but the rights owner is also often required to make a payment to cover damages in cases of abuse.

The value of this procedure lies in the fact that useful information may be speedily obtained and such information can be used later in both domestic and foreign proceedings. The descriptive seizure cannot, however, be ordered for premises outside Belgian territory.

The rights owner can simultaneously obtain through this interlocutory decision a ban on any use of the alleged infringing goods. When the claim combines both descriptive and material seizure, the president will have to balance not only the interests of the parties, but also the public interest. The president may therefore – exceptionally – decide to hear the alleged infringer in order to make a more informed assessment. In this case, he will, however, inform the patent holder prior to contacting the alleged infringer. The patent holder may subsequently decide to withdraw the case, which will then remain unknown to the alleged infringer.

It should be noted that in the context of a descriptive seizure, the applicant does not have to demonstrate the urgency of the claim, as stipulated in summary proceedings.

An infringing party which is not able to defend its interest during the procedure before the president of the commercial court has the right to lodge a so-called 'third party opposition' against the *ex parte* order. Such third-party opposition must be filed within one month of service of the descriptive seizure order, which normally occurs at the moment of the execution of the descriptive seizure. The case will be brought before the same judge who pronounced on the descriptive seizure order.

2.2 Unilateral summary proceedings

The rights holder can obtain a preliminary injunction within the framework of this unilateral procedure before the president of the commercial court, provided that extreme urgency is demonstrated by the claimant and accepted by the court. The granting of a preliminary injunction is conditional upon an assessment of the reasonable character of the case, the absence of abuse, the appropriateness of the requested measures, and the interests of the parties and the public. In this case, the rights holder can obtain a wide range of measures, within a relatively short timeframe. These include the appointment of a custodian, the execution of an expert inquiry, a prohibition of sales and a product recall.

2.3 Summary proceedings

In this procedure, the patent holder can obtain a provisional ruling, before the president of the commercial court, as to the validity of the patent and provisional prohibition measures in anticipation of a judgment in the procedure on the merits. Both parties can also request the appointment of an expert, to provide an opinion on the validity of the invention. This opinion will provide important guidance for the judge. The patent holder must demonstrate, however, the prima facie validity of the patent and the urgency of the measures claimed. A procedure on the merits, moreover, must be initiated within 20 working days – unless the court sets out another delay – following the date of the order of the judge in summary proceedings. In principle, the courts reject non-infringements actions within the framework of summary proceedings. Given the contradictory character of this procedure, rights holders tend to prefer to initiate a descriptive seizure or unilateral summary proceedings.

2.4 The cease-and-desist action

A cease-and-desist action enables the rights holder to obtain from the president of the commercial court, within a relatively short timeframe, a decision on the merits concerning the validity of the patent – if this patent is protected within Belgian jurisdiction – and prohibition measures (ie an order to cease the infringing acts). Such order can be linked to the payment of penalties in the event of non-compliance. The president of the commercial court can, furthermore, order all accompanying measures which contribute to the cessation of the infringement. At the request of the defending party, the president can also pronounce the revocation and the annulment of the patent. This procedure can be combined with a descriptive seizure. An important difference from the summary proceedings is the fact that the rights holder does not have to demonstrate urgency. However, in contrast with the position in the procedure on the merits, the judge cannot impose damages in a cease-and-desist action or a recall of the products already distributed.

2.5 Judgment within the framework of the procedure on the merits

The court may issue a judgment on the merits concerning the validity and/or the infringement of the patent. Both parties can request the appointment of an expert, to provide advice on the validity of the invention. This advice will provide important guidance for the judge. The court can pronounce:
- a cessation order against the infringing party and against the intermediary whose services are used to commit the infringing acts;
- the transfer of all useful information concerning the infringement;
- the recall and destruction of the distributed infringing goods;
- the destruction or the transfer to the rights holder of the materials and

equipment used in the production of the infringing goods;

- publication of the judgment;
- compensation for the damage to the interests of the rights holder;
- payment of the costs of the proceedings and the attorney's fees (up to a maximum level);
- the (partial) revocation and annulment of the patent. (If bad faith is demonstrated, the infringing party can furthermore be ordered to transfer the turnover realised on the infringing goods); and
- the sequestration of the counterfeit goods and of the materials and equipment used in the production of the infringing products.

Finally, on a request of the rights holder which demonstrates circumstances which jeopardise the possibility of obtaining damages, the judge can order seizure of all goods of the infringing party in the event of an infringement of an intellectual property right on a commercial scale. The defendant can initiate invalidity, non-infringement and/or revocation actions.

2.6   Final remarks

In a number of proceedings mentioned above, one has to demonstrate the urgency and the (prima facie) validity of the IP right. The Belgian judicial system does not, however, require any other special steps that must be undertaken before proceedings can be commenced.

Warning letters are not required, since the writ of summons applies as proof of default; but patent holders do send them in order to demonstrate the bad faith of the infringing party.

3.   Procedure and timescale of proceedings

In Belgium, the typical timetable for procedures on the merits, which relates both to infringement actions and declaratory actions, consists of the following steps:

3.1   At first instance

- The claim is started by formal notification of writ of summons.
- An introductory hearing takes place within eight days[1] following notification of writ of summons,
- There is an exchange of written pleadings between the parties (minimum two and a half months).
- The case is pleaded at an oral hearing before the commercial court. (Depending on the court, the date of the oral hearings might be fixed up to one year after the submission of the last written pleadings.)
- The court in principle pronounces a judgment within one month of the oral hearings.

---

[1] This delay of eight days is prolonged by:
- 15 days if the defendant is established in a country neighbouring Belgium or in the United Kingdom;
- 30 days if the defendant is established in another European country;
- 80 days in the case of any other country.

The procedure at first instance might take one year up to two years.

3.2 On appeal

- The appeal must be lodged by means of a petition within one month of the formal notification of the judgment of the court of first instance.
- The parties exchange written pleadings (minimum two and a half months).
- The case is pleaded at an oral hearing before the court of appeal. (Depending on the court, the date of the oral hearing might be fixed at up to two years after the last written pleadings.)
- The court in principle pronounces a judgment within one month of the oral hearings.

The procedure in appeal can last up to a year or more.

There is a further possibility of appeal before the Belgian Supreme Court (*Court de Cassation*) within three months of the formal notification of the decision.

3.3 Accelerated proceedings

The Belgian procedure provides for the following accelerated proceedings.

(a) *Unilateral summary proceedings*

The rights holder can obtain a preliminary injunction, within the framework of this unilateral procedure, before the president of the commercial court provided extreme urgency is demonstrated by the claimant and accepted by the court. The granting of a preliminary injunction is conditional on an assessment of the reasonable character of the case, the absence of abuse, the appropriateness of the requested measures and the interests of the parties and the public. In this case, the rights holder can obtain a wide range of measures, within a relatively short timeframe, including the appointment of a custodian, the preparation of an expert enquiry and a prohibition of sales. At the request of a rights holder who is able to demonstrate that there are circumstances which jeopardise the possibility of obtaining damages, the judge can order the seizure of all goods of the infringing party, in the event of an infringement of an intellectual property right on a commercial scale.

(b) *Summary proceedings*

See Section 2.3 above.

(c) *The cease-and-desist action*

See Section 2.4 above.

The Belgian courts are free to assess the potential relevance of EPO proceedings, but are reluctant to stay proceedings in anticipation of a decision of the European Patent Office.

4. Availability of pre-action evidence gathering/disclosure

Under Belgian procedural law, there is no pre-action evidence gathering/disclosure as in the United States and the United Kingdom, where both parties can be obliged to provide all information which they intend to rely on. There are, however, a number of procedures which allow the parties to gather information in relation to the alleged infringement prior to the actual procedure on the merits, as discussed below.

4.1 The descriptive seizure

A descriptive seizure enables a patent holder to obtain – in principle on a unilateral basis – from the president of the commercial court an order which allows the appointment of a judicial expert who will be empowered to enter the business premises of the alleged infringer in order to prepare a description of apparatus, processes, products and documentation that are "likely to establish the alleged infringement". In principle, the report of the judicial expert should not provide advice as to the validity of the patent, or on the issue of infringement. Within one month of the submission of the expert's report, the rights owner must launch a procedure on the merits relating to the alleged infringement. The president of the commercial court will grant an order on the basis of a prima facie valid IP title, together with some proof of a risk of infringement. Penalties are generally included in the order in the event of non-compliance, but the rights owner is also often required to make a payment to cover damages in cases of abuse.

The value of this remedy lies in the fact that useful information can be obtained by means of a quick and efficient procedure and the information can be used later in both domestic and foreign proceedings. The descriptive seizure cannot, however, be ordered for premises beyond Belgian territory.

An infringing party, which was not able to defend its interest during the procedure before the president of the commercial court, has the right to lodge a so-called 'third party opposition' against the *ex parte* order. Such a third-party opposition must be filed within one month of service of the descriptive seizure order, which normally occurs at the moment of the execution of the descriptive seizure. The case will be brought before the same judge who pronounced on the descriptive seizure order.

4.2 Unilateral summary proceedings

See above Section 3.3 (a).

4.3 Summary proceedings

See Section 2.3 above.

On the basis of article 877 of the Judicial Code, a party can finally request the court to order the transfer of documents containing evidence of a relevant fact, if there are significant, particular and corresponding presumptions that the other party keeps this document in its portfolio.

There is no obligation of pre-action evidence gathering/disclosure prior to the initiation of an infringement case on the merits.

5. Availability of interim relief

The patent holder can obtain interim relief within the framework of (unilateral) summary proceedings and in the context of a descriptive seizure. The different criteria in these procedures can be summarised as set out in the table below.

Procedure	Conditions	Measures	Nature
Unilateral summary proceedings	Extreme urgency. Reasonable character of the case. Absence of abuse Appropriateness of the requested measures.	Prohibition measures Description by expert	Unilateral
Summary proceedings	Urgency Prima facie validity	Prohibition measures Description by expert	Contradictory
Descriptive seizure	No urgency required. Prima facie validity. Reasonable indication or suspicion of infringement. Only in cases of seizure: infringement cannot be reasonably disputed andthe requested measures are reasonable.	Description Seizure Ban on the disposal of alleged infringing goods	In principle unilateral

Belgian procedural law does not contain specific provisions for protective letters. There is thus a discussion in Belgian doctrine on whether the use of protective letters is allowed in Belgium. It seems doubtful whether the courts will take these protective letters into account.

6. Disclosure

Belgian procedural law does not provide for disclosure as in the United States and the United Kingdom, where both parties can be obliged to provide all information on which they intend to rely.

The following proceedings – all of which must be initiated by the applicant – allow the parties to gather information about the counterparty prior to the actual procedure on the merits:

- descriptive seizure (see Section 2.1 above);
- unilateral summary proceedings (see Section 3.3 (a) above); and
- summary proceedings (see Section 2.3 above).

On the basis of article 877 of the Judicial Code, a party can finally request the court to order the transfer of documents which contain evidence of a relevant fact, if there are significant, particular and corresponding presumptions that the other party keeps this document in portfolio.

In principle there is no limitation on the use of the documents disclosed. The information obtained in the framework of the descriptive seizure can, however, no longer be used if the applicant fails to initiate proceedings on the merits within one month of the expert drafting his report on the descriptive seizure.

All court hearings regarding patent litigation are public. The parties can, however, request the court not to reproduce company or patent information in its decision. The judge can make a decision on this at his discretion.

7. Evidence

Under Belgian procedural law, written evidence prevails. The parties can invoke documentary evidence, such as foreign judgments, written arguments and experts' reports. In principle, no witness statements are used.

In disputes between commercial enterprises, witness statements are available in theory. Nonetheless, Belgian commercial courts have shown great reluctance to allow witness statements in commercial disputes, because the procedure is considered time consuming and arduous, and the use of written documents has become the norm.

It should also be noted that in contrast with the Netherlands, France, Germany or Austria, under Belgian law (as in Italy) a party is not allowed to offer its own testimony as proof. Belgian law does not allow party witnesses.

An expert is appointed by the court. The parties can hire experts to draft reports, which can be submitted as evidence. Exceptionally, the expert is heard by the court.

Belgian procedural law does not allow cross-examination. All questions to possible witnesses (which, as noted above, are exceptional) must be addressed to the court, which will hear the witnesses.

8. Law

The Belgian courts take steps to determine whether there is a patent infringement. Belgian patent law recognises both *direct* and *indirect* infringement.

There are two forms of direct infringement:

- literal infringement; and
- non-literal infringement (also often called infringement by equivalence).

Literal infringement occurs where all elements of the claim are infringed by the alleged infringer. *Non-literal infringement* exists where not all elements of the claim are literally infringed by the alleged infringer, but where the allegedly infringing product or process can be seen to be equivalent to the patented product or process.

The Belgian Patent Act is silent as regards the doctrine of equivalence. However, Belgian case law does apply the principle of equivalence, but cautiously and within certain limits. There are only a limited number of cases dealing with the issue of equivalence and the courts have not defined a clear and single equivalence test. However, despite these uncertainties there are certain principles which are often used and applied by Belgian courts in evaluating non-literal infringement.[2]

A first standard which has been suggested in the earlier literature is to establish equivalence by drawing a distinction between 'technical functions' and 'technical factors', known herein as the 'technical functions/features test'. Technical factors are

---

2   See R Peeters and B Vandermeulen, *De equivalentieleer in de Belgische octrooirechtspraak*, IRDI 2003, p 131.

the concrete, physical means, while technical functions relate to the function of these factors (ie their functioning or activity and interaction). Under this test, there is infringement by equivalence of technical functions if the technical factors in the allegedly infringing product realise the same technical function as those of the patented product. This theory is not easy to apply, however, and remains somewhat abstract and difficult to comprehend.

In more recent cases, various other principles have been applied by the Belgian courts, which have brought them across from established equivalence doctrines in other countries, including the Netherlands. Establishing equivalence under recent case law consists of various steps:

- An evaluation of equivalence starts by looking at the so-called 'essential elements' of the invention. Even though it is not entirely clear what exactly is meant by 'essential elements', it definitely contains all elements which form a contribution to the art;
- Once all essential elements of the invention have been established, the issue of whether the allegedly infringing product infringes the patent is examined. It is here that various tests for equivalence may be applied by Belgian courts:
  - One of these standards is the so-called 'function-way-result test'. Under this test, a product or process infringes a patent if the allegedly infringing product or process performs substantially the same function in *substantially the same way* to achieve substantially the same result; and
  - Another test which has been applied by Belgian courts is the so-called 'insubstantial differences test'. Under this test, the allegedly infringing product or process infringes the patent if the differences are insubstantial.

As in the European Patent Convention, article 5, section 2 of the Belgian Patent Act provides that "an invention shall be considered to be new if it does not form part of the state of the art". The state of the art is defined as "everything made available to the public by means of a written or oral description, by use, or in any other way, before the date of filing of the patent application" and comprises the content of all Belgian, European and PCT (Patent Cooperation Treaty) patent applications – with effect in Belgium – submitted prior to the date of filing of the relevant patent application. The Belgian patent system furthermore does not provide any geographical limitation on the place of disclosure of the prior documents. It also provides that a document can only be detrimental to the novelty of a patent application if it contains all essential characteristics of the invention and results in the same technical effect. The invention furthermore must result directly and unambiguously from the document which is claimed to be detrimental to the novelty. The disclosure of a prior document will not be detrimental to the novelty of the invention if it has been made available within six months prior to the date of filing of the patent application and:

- the disclosure is the direct or indirect result of an abuse against the applicant; or
- the document was made available by the inventor at an officially recognised exhibition.

The issue of *sufficiency* is dealt with under article 17(1) of the Belgian Patent Act (BPA) and article 9, section 1, 6° of the Royal Decree of December 2 1986. The wording of article 17(1) BPA is identical to Article 83 EPC (ie an application shall disclose the invention in a manner sufficiently clear and complete for it to be carried out by a person skilled in the art). The wording of article 9 section 1, 6° of the Royal Decree of December 2 1986 is identical to rule 27(e) EPC 1973 (rule 43(e) EPC2000) (ie the application describes in detail at least one way of carrying out the invention claimed, using examples where appropriate and referring to the drawings, if any).

In general, Belgian courts will follow EPO practice regarding sufficiency of disclosure (enablement). According to this practice, enablement should be determined at the relevant date (eg the priority or filing date). In general, deficiencies cannot be cured later. An important aspect of the enablement requirements is that the claims should be enabled over the whole application. It has been consistent practice of the boards of appeal that sufficiency of disclosure within the meaning of Article 83 EPC must be assessed on the basis of the application (patent) as a whole, including the description and claims as well as the common general knowledge of a person skilled in the art. It is commonly accepted that an invention is in principle sufficiently disclosed if at least one way is clearly indicated to enable a person skilled in the art to carry out the invention. Nevertheless, more technical details and more than one example may be necessary in order to support claims of a sufficiently broad scope. Notably, in *inter partes* proceedings, the opponent has the onus of demonstrating on the balance of probabilities that a claim is not sufficiently made out (ie over the whole scope of the claims). The disclosure is aimed at the person skilled in the art, who may use his common general knowledge to supplement the information in the application (patent).

9. Available remedies

The available remedies differ, depending on the relevant procedure.

9.1 Descriptive procedure

Within the framework of a descriptive seizure, it is possible to obtain:

- the appointment of a judicial expert, who will be empowered to enter the business premises of the alleged infringer in order to prepare a description of apparatus, processes, products and documentation that are "likely to establish the alleged infringement"; and
- a ban on any disposal of the alleged infringing goods.

9.2 Unilateral summary proceedings

In the context of unilateral summary proceedings, it is possible to obtain:

- the appointment of a custodian;
- the execution of an expert inquiry;
- a prohibition of sales;
- the seizure of all goods of the infringing party; and
- a product recall.

9.3 Summary proceedings

The following remedies are available in summary proceedings:

- a provisional ruling on the validity of the patent;
- provisional prohibition measures;
- the appointment of an expert to draft a report on the validity of the patent and on the infringement; and
- a product recall.

9.4 Cease-and-desist action

In a cease-and-desist action, the following remedies are available:

- a decision on the merits on the validity of the patent;
- a cessation order;
- all accompanying measures which contribute to the cessation of the infringement; and
- revocation and the annulment of the patent.

9.5 Procedure on the merits

In the framework of a procedure on the merits, it is possible to obtain:

- a cessation order against the infringing party and against the intermediary whose services are used to commit the infringing acts;
- the transfer of all useful information concerning the infringement;
- the recall and the destruction of the distributed infringing goods;
- the destruction or the surrender to the rights holder of the materials and equipment used in the production of the infringing goods;
- the publication of the judgment;
- compensation for the loss suffered by the rights holder;
- payment of the proceeding cost and the attorney's fees (to a certain maximum amount);
- revocation and the annulment of the patent; and
- the seizure of all goods – including bank accounts – of the infringing party.

If the rights holder can demonstrate the bad faith of the infringing party, he can also claim:

- the transfer of the turnover realised by the infringing party on the infringing goods; and
- sequestration of the counterfeiting goods and of the materials and equipment used in the production of the infringing products.

The EU Enforcement Directive 2004/48/EC has been implemented in Belgium via three Acts of May 9 and 10 2007. Although most enforcement measures were already provided for in Belgian procedural law, some new measures have been introduced, such as seizure of the infringing party's bank accounts. The conditions under which the measures can be requested have also been brought in line with the directive and there has been a harmonisation of the applicable measures for the different intellectual property rights.

In order to calculate financial remedies, the judge will take into account the negative economic consequences for the patent holder and the illegitimate profits which the infringing party has enjoyed. The negative economic consequences for the patent holder include:

- loss of profits of the rights holder;
- devaluation of the exclusive character of the patent;
- damage to the reputation of the patent holder and its products; and
- administrative and judicial costs related to the tracing, examination and prosecution of the infringements.

Compensation will, in principle, be calculated on the basis of loss of profits of the rights holder and the illegitimate profits enjoyed by the infringing party. If this is not possible, the judge will grant a fixed amount, based on an *ex aequo et bono* estimation of the damage. In this estimation, he can take into account royalties which have actually been paid, or a reasonable estimation of royalties which would have been paid if the infringing party had obtained a licence.

The damages are calculated at trial, but the courts sometimes decide to rule first on validity and infringement and to stay proceedings regarding damages until a judicial expert has assessed the damages incurred.

The effect of an appeal on the enforcement of the decision depends on whether the court has held that the decision at first instance is provisionally enforceable. If the court has indeed held that this is so, an appeal against the decision does not suspend its execution.

10. Costs

The costs of an average patent action can be estimated at approximately €100,000 up to €250,000.

In line with the implementation of Article 14 of the EU Enforcement Directive, the Belgian judicial system allows the winner to recover part of the judicial costs, including the attorney's fees. The amount of the recovery of the judicial costs is stipulated in a royal decree and depends on the value of the claims. As compensation for the judicial costs of the winner, the judge can order a payment of up to €14,000 (for claims of €250,000 to €500,000), €20,000 (for claims of €500,000 to €1,000,000) and €30,000 (in claims of more than €1,000,000).

Denmark

Peter-Ulrik Plesner
Sture Rygaard
Caroline Thufason
Plesner

1. Lawyers

In patent litigation in Denmark, the team normally comprises lawyers, both internal and external, patent attorneys and people from the company responsible for the legal and/or technical side of the relevant patents. All qualified lawyers who have passed their bar exams can appear before the Bailiff's Courts.

Lawyers are not limited to their local court, but can conduct a case in any court of the level before which they are authorised to appear. In Denmark, practitioners must pass a bar exam to appear before the High Court and the Maritime and Commercial Court and another exam to appear before the Supreme Court.

Figure 1: The court system

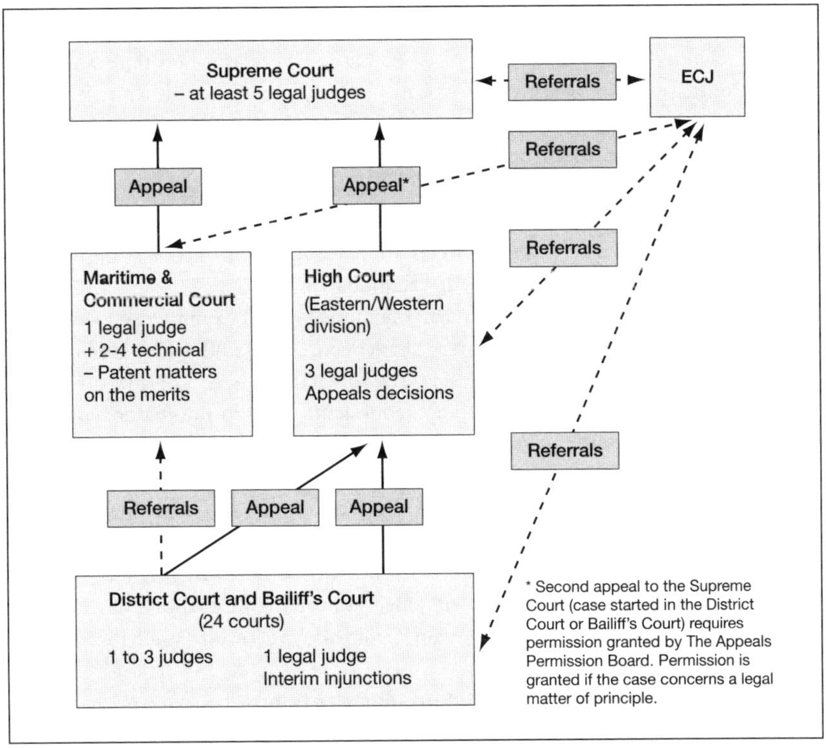

2. The court system

2.1 Overview of the civil court system

Danish civil procedure is subject to rules set out in the Administration of Justice Act. The Act contains rules regarding organisation of the courts (see Figure 1), as well as for the preparation of cases before a court and court hearings.

The ordinary courts are organised in a three-level hierarchy consisting of the district courts, the High Courts and the Supreme Court. The 24 district courts are located around the country, whereas the two High Courts are organised in two divisions, The Western Division covering Jutland situated at Viborg and the Eastern Division situated in Copenhagen covering the rest of Denmark, while the Supreme Court has overall jurisdiction and is located in Copenhagen.

In general, a case may be tried twice in the Danish court system, referred to as the principle of two instances. Appeal from a district court lies to the High Courts, while appeal of cases commenced before the High Court lies to the Supreme Court. However, depending on the circumstances, permission may be granted by the Danish Appeals Permission Board to bring a case before the Supreme Court as third instance where such case deals with an important issue of principle.

Patent matters on the merits are dealt with by the Maritime and Commercial Court in the first instance. The Maritime and Commercial Court is located in Copenhagen and is considered one of the ordinary courts, although it only hears cases concerning international trade, maritime cases, bankruptcy cases, competition cases and cases relating to intellectual property rights (except for copyright), including cases concerning patents. Cases before the Maritime and Commercial Court are heard by either three or five judges, one legal judge and either two or four expert lay judges. The lay judges are appointed from a permanent panel of experts on a case-by-case basis to ensure that those appointed are qualified in the field of the specific proceedings, depending on whether it is a technical, scientific or pharmaceutical case. Judgments rendered by the Maritime and Commercial Court can be appealed as of right directly to the Supreme Court, and the Supreme Court will hear a full trial (ie of both facts and law).

Apart from cases dealt with by the Maritime and Commercial Court, as a general rule civil cases are heard before a district court unless otherwise specifically stated in the Administration of Justice Act. The district courts also perform the functions of Bailiff's Court and Notary Public. Requests for interim injunctions are submitted to the Bailiff's Court and are heard by a single legal judge. The High Court hears appeal cases from the district courts and cases referred from the district court, where they are regarded as having general public importance, or at the request of a party. The Supreme Court hears appeal cases from the High Court and from the Maritime and Commercial Court. In the district court a case is heard by one judge, but the judge may be assisted by two expert lay judges. In the High Court cases are always heard by at least three judges, while Supreme Court cases are heard by at least five legal judges.

2.2 Decisions from other jurisdictions and the Danish courts

In accordance with Denmark's obligations pursuant to its EC membership, the

Danish courts take note of decisions from the European Court of Justice. Decisions from other jurisdictions concerning international conventions or treaties acceded to by Denmark can, of course, also be of interest in Denmark. As regards patent litigation, the courts probably take particular note of decisions from the other Nordic countries, as the Danish Patent Act is based on a joint Nordic commission report. For patents and the assessment of validity, novelty, inventive step and so on, the Danish courts will generally follow such principles and interpretation as can be deduced from the decisions of the European Patent Office (EPO) and the Technical Board of Appeal.

2.3 Patent-related actions in Denmark and pre-action requirements

It is possible to file a declaration for non-infringement suits and a revocation action. A request for revocation can also be filed administratively with the Danish Patent and Trademark Office. The available actions are discussed at Section 9 below. There are no special requirements or steps to be fulfilled before action before the court can be commenced, but the suit has to relate to an actual dispute (see Section 11 below).

Warning letters are not required, but are generally considered a good idea. The effect and intention of warning letters are discussed further at Section 11 below.

3. Procedure and timescale of proceedings

3.1 Preparation and conduct of hearings

In Denmark there is no difference between procedures for infringement actions and those for declaratory actions, and there is no pre-trial preparation.

All patent trials in Denmark consist of a preparatory phase and an oral hearing. The preparatory phase is conducted in writing, while the hearing of the case is oral.

The plaintiff institutes an ordinary patent infringement action by filing a writ of summons with the Maritime and Commercial Court in Copenhagen. The writ of summons must list the claims submitted by the plaintiff together with the allegations supporting the claims, and must include a short presentation of the facts together with the documents and other evidence the plaintiff wishes to invoke in support of the claims. After the writ of summons has been duly served on the defendant, the defendant files his statement of defence. The written preparatory phase will include further exchanges of at least one more written pleading from each party in relation both to the facts and to the law. In patent cases, during the preparatory phase, at least one of the parties will normally ask the court to appoint one or more technical experts in order to obtain an expert appraisal of the technical aspects of the case on the basis of written questions from the parties. When the experts have submitted their appraisal to the court, the parties will normally submit another written pleading addressing the findings of the expert appraisal. Then the case is normally ready for the oral hearing. In this phase the parties specify their claims and allegations and the evidence upon which the court is to base its decision during the oral hearing.

New claims, allegations and evidence may be submitted or altered freely until the end of the written preparation (ie until the oral hearing is scheduled).

The oral hearing of the case is initiated by the parties stating their claims. The plaintiff then orally presents the factual issues of the case to the court. Usually, this is done by means of a chronological review of the case. At the same time, the relevant parts of the invoked documents are recited to the court. The plaintiff's presentation should be objective and should also include the documents and so on which the defendant intends to invoke. After this, the defendant can make additional comments. Then the parties produce their evidence and each party argues its case. The pleadings include interpretation of the case, and reference to the relevant law and judgments which may support the party's allegations. After the oral hearing has closed, the court may render its judgment immediately in the same court meeting, or the case may be set down for judgment.

In patent infringement proceedings on the merits, the defendant may claim that the patent-in-suit is invalid, and the court will decide on such a claim during the proceedings.

It is possible to have accelerated proceedings in Denmark, as cases can be heard under the 'fast track' procedure at the Maritime and Commercial Court. This requires that the preparatory phase is prepared quickly, with few pleadings and is scheduled for trial as soon as possible. The deadlines in the preparatory phase are accelerated by the court. The fast track can be initiated either by attorneys, or at the assessment of the court. However, in patent matters fast-track proceedings will rarely be used in practice, owing to the complexity of patent-related issues. For this reason, interim injunctions are often used in patent matters (see Section 5 below).

3.2 The potential effect of EPO opposition proceedings

The fact that the relevant patent is subject to pending nullity proceedings or opposition proceedings by the EPO and the Technical Board of Appeal does not automatically prevent an ordinary patent case from being stayed. However, ordinary patent proceedings can be conducted according to Danish case law depending on the circumstances. This was the case in U. 2001.1535 H (*Danish Weekly Law Journal* 2001, page 1535), where the Supreme Court allowed a case to be stayed pending a decision from the EPO opposition division, since the decision would without doubt have had an impact on the case before the High Court (see Section 345 of the Administration of Justice Act).

There is no general court practice showing whether patent trials will generally be stayed pending EPO opposition division or Technical Board of Appeal decisions, or in relation to limitation requests filed by the proprietor after the European Patent Convention (EPC) 2000 opened that possibility.

In Section 53(a) and (b) of the Danish Patents Act, it is stated that administrative hearings of a patent's validity cannot take place at the same time as a trial on the same patent. The Patent Authority thus has to reject a request for re-examination by a third party while revocation proceedings are taking place. The Patent Authority is only to suspend the re-examination of the patent on request by a third party and not by the patent proprietor himself. A final decision is made by the Patent Authority if requested by the patent proprietor, even in the event of a suit.

Pending nullity proceedings or opposition proceedings do not prevent an interim injunction, which can be initiated as long as the patent has not been claimed invalid. It appears from Danish case law that an interim injunction has been granted even though the patent in suit had been invalidated in both the United Kingdom and in Germany.

The consequence of pending nullity proceedings in Denmark may be that the case is stayed upon receipt of the relevant decision.

4. Availability of pre-action evidence gathering/disclosure

Under Danish law there are few means of gathering pre-trial evidence. The parties have the option of obtaining an extrajudicial expert appraisal, but this is rarely used in patent litigation. The most common way of obtaining pre-trial evidence is to gather it in accordance with the rules on preservation of evidence.

4.1 Preservation of evidence

The Danish legislation on preservation of evidence of infringement of patents was introduced in recognition of the difficulties which rights holders would often encounter due to lack of evidence when trying to obtain adequate damages for infringements. This provisional measure entitles the claimant to investigate an alleged infringer's business premises in order to secure evidence of an infringement of the claimant's intellectual property rights (see the Administration of Justice Act, Chapter 57a, now Consolidated Act No 1261 of October 23 2007 with subsequent amendments).

The rules have now been in force for more than seven years and practice shows that they are widely used and an effective means for the rights holders to establish the extent of any infringements. This is very useful as evidence in any subsequent action for damages and may help the rights holder to identify the sources and buyers of infringing products.

4.2 Procedure and criteria for granting preservation of evidence

Following a request from the owner of intellectual property rights or a person entitled to take legal proceedings, the Bailiff's Court may order necessary measures to be carried out on the premises of an alleged infringer, in order to preserve evidence of an infringement of intellectual property rights and/or to ascertain the extent thereof. In order to initiate the procedure the claimant must show (see Section 653(1) of the Administration of Justice Act) that there is at least a 50% likelihood that:

- the defendant has infringed the rights in question in the course of trade; and
- there is reason to believe that evidence of the alleged infringement and its possible extent can be found at the stated premises.

Preservation-of-evidence actions have been shown to be an effective means of obtaining evidence in Denmark and, in contrast with some jurisdictions, a copy of the relevant data and documents is provided to the rights holder. The investigation of the infringement may include all information which could be considered of importance in order to establish whether an infringement has taken or will take place, as well as the extent of such an infringement, including products for sale, machines and other manufacturing equipment, accounting material, invoices, order forms, marketing material and other documents, information stored on computers, and software and electronic storing facilities (see Section 653(3) of the Administration of Justice Act).

The investigation can take place at any premises which are located in Denmark and which are owned by the alleged infringer, or which the alleged infringer has at its disposal. A request for preservation of evidence cannot be directed against a third party (eg the Danish Medicines Agency).

The main rule is that the application for preservation of evidence must be filed with the Bailiff's Court in the judicial district of the alleged infringer's premises. Searching the alleged infringer's premises is not always an adequate measure, since products and data are sometimes stored elsewhere in another judicial district. If it turns out, for example, that the infringing products are stored at another address in another judicial district, the Bailiff's Court can continue the investigation at that address. Preservation of evidence may be made at all premises at the defendant's disposal, and therefore the possibility of extending the search for evidence to a third party's premises is allowed in some cases (eg if a third party is storing digital media or data on a server on behalf of the infringer). The court may also order the defendant's accountant to surrender the defendant's accounts.

The Bailiff's Court will request the claimant to provide security in order to cover any damages and expenses of the defendant in the event that there is no proof of infringement, unless the infringement has been substantiated and not just rendered probable.

The court may commence the investigation without notifying the defendant, in circumstances where such notification might be expected to involve a risk that evidence will be removed, destroyed or altered. In practice, prior notification to the defendant is usually dispensed with. Upon arrival at the defendant's premises, the Bailiff's Court will then first inform the defendant of his right to have an attorney present and the court will, at the defendant's request, suspend the investigation for about an hour until the attorney has arrived.

If there is a presumption that the investigation will cause the defendant damage or inconvenience (taking into consideration the defendant's right of privacy, trade secrets or other considerations), which is disproportionate to the claimant's interest in the investigation, the court may dismiss the claimant's request or limit the scope of the investigation (see Section 653(4) of the Administration of Justice Act). It is possible for the court to decide that the claimant and/or the claimant's representatives are not to be granted access or to grant only limited access to the defendant's premises during the investigation (see Section 653b(3) of the Administration of Justice Act). It is also possible for the court to request assistance from an independent technical expert in connection with the investigation (eg a computer expert and/or an expert of the relevant technical field will often assist the judge upon the claimant's request).

The investigation cannot include material which contains information on circumstances as to which the defendant is excluded from giving testimony due to a duty of confidentiality (eg for ministers, doctors and attorneys – see Section 653(5) of the Administration of Justice Act). However, this limitation on the possibility of preserving evidence does not apply in relation to material which may put the witness or his close relatives at the risk of liability (see the witness exemption rules in Section 171 of the Administration of Justice Act).

In relation to balancing the claimant's interest in the preservation of evidence and the defendant's interest in protecting its trade secrets, it is stated in the preamble to the Administration of Justice Act that the proprietor of a process patent has a particularly strong interest in the preservation of evidence if shifting of the onus of proof in relation to process patents for new compounds in the Danish Patents Act does not apply. Hence, in this situation there are very strong reasons for the court to follow a request for preservation of evidence, even though the preservation of evidence may result in disclosure of the defendant's trade secrets.

As the rules on preservation of evidence in relation to process patents have only recently come into force, there is not yet any relevant case law from the Danish courts. In applying the rules for preservation of evidence, it will also be important to establish to what extent the claim for infringement can be substantiated (ie the stronger the (indirect) evidence of the alleged infringement is, the greater are the chances of the preservation of evidence being granted).

The claimant must bring a confirmatory action against the defendant within four weeks of the Bailiff's Court's notice that the investigation has been finalised. If the defendant has not waived prosecution, and should the claimant fail to bring action in time, the Bailiff's Court and the claimant must return the objects and documents seized, copies produced and other evidence found during the investigation, and any evidence of infringements obtained during the preservation will be inadmissible.

It is common for an application for preservation of evidence to be heard together with an application for an interim injunction so that both enforcement proceedings may take place in a single action on the same day.

Under the Danish Administration of Justice Act there are no rules on confidentiality to the effect that evidence gathered according to the rules on preservation of evidence should not be exported to other jurisdictions, or that the information should be kept confidential.

The procedure of preservation of evidence does not oblige the parties to commence substantive proceedings, but the party requesting the procedure has to file a confirmatory action in order to be able to use the evidence at a later date at trial. It is not a condition for the procedure of preservation of evidence that the party should have filed a writ of summons concerning infringement. If the results are exported to another jurisdiction, there is no obligation to commence substantive proceedings there.

5. Availability of interim relief (especially injunctions)

5.1 Initiation of proceedings for interim relief

Interim injunctions are the most commonly applied preliminary relief. A plaintiff institutes an interim injunction action by filing a request for grant of interim injunction with the Bailiff's Court in the jurisdiction of the defendant's domicile. There are no specialised courts for intellectual property disputes in relation to interim injunctions. After being summoned to appear in the enforcement proceedings, the defendant submits its statement of defence. The parties normally exchange two further written pleadings before the pre-trial work is completed. The legal judge will not appoint any experts in order to obtain an appraisal of the technical aspects of the case. Instead, the parties are allowed to submit unilaterally procured expert appraisals. Either party may appeal the decision of the single judge to the High Court.

5.2 The criteria applied by the courts for the granting of injunctions in patent actions

The rules on injunctions are stated in the Danish Administration of Justice Act, Chapter 57 on prohibition. The judge can grant an interim injunction if the plaintiff can establish *prima facie*, or show that it is probable, that the following criteria are met:

- the actions at which the injunction is directed infringe the rights of the plaintiff;
- the defendant would perform the actions against which the injunction is directed; and
- the purpose of the application would be forfeited if the plaintiff had to resort to ordinary court proceedings.

The Bailiff's Court may refuse to grant an injunction if it will cause the defendant harm or inconvenience which is disproportionate to the plaintiff's interest in having an injunction granted. This "principle of proportionality" requires the court to consider the balance of convenience of the parties.

The Bailiff's Court make the injunction conditional on the plaintiff providing security in the form of a bond, unless the infringement is considered substantiated and not just shown to be probable. The size of the bond should reflect the loss the defendant could suffer in the event that the injunction is cancelled or found unjust. The interim injunction will not be issued until the security has been provided.

In the event that the judge grants an interim injunction against the defendant, the plaintiff must bring an action before the Maritime and Commercial Court for a permanent injunction within two weeks of the granting of the interim injunction. The parties are entitled to appeal the judgment of the Maritime and Commercial Court to the Supreme Court.

In interim injunction proceedings, the plaintiff cannot make any claim for damages, as this issue is dealt with in the following action for a permanent injunction, if an injunction is granted.

The plaintiff is only required to pay a court fee of Dkr300 in connection with the institution of the interim injunction action, but the plaintiff will be required to pay the normal court fee of 1.2% of the subject matter of the case (as well as an additional 1.2% when the case is scheduled for trial) in connection with the following action for a permanent injunction.

During interim injunction proceedings, the defendant may also argue that the patent-in-suit is invalid. In such a situation, the judge does not examine whether the invention which led to the patent-in-suit involved any inventive step (see Section

2(1) of the Danish Patens Act (EPC, Article 52)), but *only* refuses to grant an injunction due to the invalidity of the patent-in-suit, if the defendant is able to submit documents that have not been assessed by the relevant patent authorities (eg EPO) and that contain actual information detrimental to the *novelty* of the invention. In other words, if an issue regarding novelty (eg a publication) has been raised but rejected by the EPO during the examinations leading to the granting of the patent, the judge will most likely not consider such issues again under an interim injunction action.

5.3 Without-notice applications

In patent litigation, the defendant will nearly always be heard before the court decides whether to issue an injunction. However, in special circumstances, the Bailiff's Court is empowered to omit notifying the defendant, but this possibility is only rarely used (eg in piracy cases). Danish case law provides another option where the defendant is not heard. When preservation of evidence is combined with an injunction in a joint case, the rules of preservation of evidence imply that the defendant is not aware that the judge will visit the defendant's address to preserve evidence, and it can be advantageous in some cases to have a joint case processing both aspects of the case concurrently.

Interim relief is only available where an applicant cannot wait for the result of an ordinary trial. If a warning letter or another action which shows knowledge of infringing activity is not followed by further action within reasonable time, the patent proprietor may risk not being able to enforce his intellectual property (IP) rights by means of an interim injunction – because one of the conditions for an interim injunction in Denmark is that enforcement should not await an ordinary trial. Passivity/acquiescence may also result in damages awards being reduced, or forfeiture of the right to damages.

In this context, Danish law does not have any fixed term after which relief is not available. However, recent case law has shown that inactivity for one year will not necessarily prevent a plaintiff from obtaining injunctive relief. Injunctive relief will normally be available in patent matters if the request is filed within three to four months and probably also somewhat later; and it will probably depend on the complexity of the matter and the need for tests and analysis to be performed prior to the request. Thus, the patent proprietor has some time to make an assessment of his rights and of what steps should be taken. One recent case has stated that delay definitely occurs with the passing of eight years in regard to injunctive relief. However, the situation is not quite settled as far as the right to claim damages in a later civil case is concerned.

5.4 Availability of protective briefs

Protective briefs are not recognised under Danish law. There is no need to file an action for a negative declaratory judgment in order to prevent an interim injunction since the filing of a claim for negative declaratory judgment does not influence whether or not an injunction will be granted. This has been confirmed in many Enforcement Court decisions.

5.5 Pros and cons of applying for injunctions

In balancing the pros and cons of applying for injunctions, the main reason for using them is the fact that they are by far the fastest way for the patent proprietor to obtain a decision in terms of discovering infringements, and may be the only way to avoid otherwise irreparable harm resulting from any infringement of patent rights.

The main disadvantages of an interim injunction are the often quite considerable security required by the Bailiff's Court, and the fact that the plaintiff is strictly liable to compensate all damage caused by a groundless interim injunction. There is no possibility for a potential counterclaim for summary judgment on revocation. The decision on revocation can only be given in an ordinary patent case on the merits, or as a counterclaim during the confirmatory suit.

6. Disclosure

6.1 When disclosure is available

In order to implement Article 8 of Directive 2004/48/EC of April 29 2004 on the enforcement of intellectual property rights – introducing an obligation for the defendant to disclose certain information upon the plaintiff's request – new rules came into force in 2006 (Act No 279 of April 5 2006). The rules are applicable only in connection with court cases on the merits involving infringement of intellectual property rights (ie not injunctions), and their purpose is to provide rights holders with more effective means for obtaining information and evidence of previous infringements, the extent of such infringements, as well as information about the parties involved. Pursuant to these rules, a party can request that the court order the defendant or a third party to disclose information in regard to goods or services that constitute an infringement if

- the party was found to be in possession of the infringing goods on a commercial scale;
- the party was found to be using the infringing services on a commercial scale;
- the party was found to be providing services used in infringing activities on a commercial scale; or
- the party was indicated by a person referred to in the three points above to be involved in the production, distribution or supply of the infringing goods or services in question.

The conditions are not cumulative. The disclosure of information includes:

- names and addresses of the manufacturers, distributors, suppliers, and other previous owners of the goods or services in question, as well as the intended wholesalers or retailers;
- information on the quantities manufactured, delivered, received or ordered and the price obtained for the goods or services in question.

So far, the interpretation of the provision by the courts has been quite expansive.

6.2   Limitations on the use of disclosed documents

The request for disclosure presupposes that disclosure of the documents in question does not lead to the disclosure of circumstances about which the party would be excluded or exempt from giving testimony as a witness, or answers which might incriminate him (see Sections 169 to 172 of the Administration of Justice Act). A witness is exempt from answering questions if the witness or his/her closest relatives are thereby exposed to punishment or loss of welfare, or if other substantial detriment is thereby inflicted on the person or his/her closest relatives. Other substantial detriment may be financial detriment, such as disclosure of trade secrets and other confidential information. On this basis, an alleged infringer will be able to avoid the negative effect that a refusal to disclose documents will be deemed prejudicial to him. The witness exemption rule also covers employees of a party who is exempt from giving testimony.

A request for disclosure can be fully or partially rejected if the disclosure will cause the defendant harm or inconvenience which is disproportionate to the plaintiff's interest in having the information disclosed. This principle of proportionality requires the court to consider the overall balance of convenience. Moreover, this right to receive information from the defendant about an infringement of intellectual property rights cannot be used by the plaintiff for the purpose of obtaining evidence to show that infringement exists.

Should the defendant or the third party refuse to comply with the request or give false information, without a valid reason for doing so, the party in question can be fined by the court and may be liable to imprisonment for a period of up to six months. The defendant's refusal may also have a prejudicial effect on his case.

The above mentioned rules apply to infringement of intellectual property rights. If disclosure is not available pursuant to these rules, the plaintiff may request disclosure pursuant to the general rules on disclosure (see Sections 298 to 300 of the Administration of Justice Act). These rules also allow for disclosure directed against third parties.

6.3   Protection of confidential information

As regards confidentiality, the evidence is normally produced in open court. It is not possible to submit documents as exhibits in a case under any terms of confidentiality. Patent litigation is open to the public in Denmark, but the judge may choose to conduct hearings behind closed doors and make the case inaccessible to the public, especially if trade secrets may be revealed. However, this does not ensure complete confidentiality as the attorneys may argue the case (which takes place in open court) on the basis of the information given during the oral testimony behind closed doors and the judge may refer to the information in his decision (which is available to the public). Moreover, the information is not exempt from private reproduction. Therefore, evidence will generally be made public if produced as evidence in trial.

7.   Evidence

In patent litigation it is important to distinguish between evidence that is put before the court in ordinary patent proceedings and that which is used in interim injunction proceedings.

7.1 Ordinary patent proceedings on the merits

An expert appraisal is often conducted in ordinary patent proceedings. It is important to note that in ordinary proceedings on the merits it is not possible for the parties to submit unilaterally procured expert statements to substantiate their claims. In such patent proceedings the Maritime and Commercial Court will, at the request of the parties, usually appoint (independent) experts suggested by the parties in common in order to obtain an appraisal of the technical aspects of the case, and the experts must provide the court with impartial technical assistance.

The number of expert witnesses varies from case to case. In some cases one expert witness is sufficient, but in other cases the court appoints two or more expert witnesses – for example, a specialist within the specific technical field and a patent attorney. The parties determine the kind of expertise needed to address the subject matter of their questions. Following such agreement, a relevant trade organisation or a university (eg the Danish Association of Patent Agents and the Danish Pharmaceutical University) will be approached by the parties and requested to suggest one or more suitable experts. The expert(s) suggested will be formally appointed by the court – provided that there are no material objections to their appointment (eg a conflict of interest). In such proceedings, the expert evidence should be procured by directing questions to the independent experts appointed by the court. Each party will suggest a number of questions to be asked, or they may agree to joint questions. The expert(s) will prepare (often jointly) written answers to the questions asked by the parties. The expert(s) will be requested to answer the questions asked by the parties on the basis of literature and other documentation, and not on the basis of speculation (eg not "What do you think persons skilled in the art thought of a certain method [X] years ago?"). The expert(s) will be requested to give oral witness statements in court in order to verify the conclusions of their appraisal. Both parties will have the opportunity to examine the experts.

7.2 Interim injunction proceedings

A very different procedure is followed in interim injunction proceedings, where the parties are allowed to present their "own" experts and will normally submit one or more unilaterally procured expert appraisals in order to substantiate the claims submitted in the proceedings. The experts will normally witness in court in order to verify the conclusions of their statements. Each party begins with a direct examination of its witnesses, after which the opponent will be given the opportunity to cross-examine the witnesses.

Unilaterally procured expert statements are normally prepared in close cooperation between the experts and the attorneys and patent attorneys for the instructing party, and therefore the statements are not considered as impartial as those from (neutral) court-appointed experts. The experts (and the parties and their attorneys) are free to determine the style and content of the statement, but normally it corresponds to the style of expert statements in ordinary patent proceedings (ie question/answer form, and the questions should only concern the technical aspects of the case).

7.3 How the courts deal with experiments

In patent litigation on the merits, the parties may ask court-appointed experts to undertake certain experiments. The parties may also ask an independent laboratory or technical facility to conduct experiments and analysis, which are also allowed to be produced in court. Furthermore, the parties may submit data from experiments which they themselves have performed. Internal data from the parties' own laboratories is equally admissible, although the weight given to results from internal experiments is not as great as those undertaken by independent experts.

In interim injunction proceedings, the parties may submit results of experiments and analysis carried out by third parties (laboratories etc) or by themselves.

In general, it can be concluded that there are very few limitations as to what evidence can be put before the court in patent litigation.

8. Law

8.1 The test for patent infringement

In Danish law, a third party infringes the patentee's effective monopoly for the commercial exploitation of his patent if:

- he makes, offers, puts on the market or uses a product which is the subject matter of the patent, or imports or stocks the product for these purposes;
- without the consent of the proprietor of the patent, he uses a process which is the subject matter of the patent or offers the process for use in Denmark if the person offering the process knows, or it is obvious in the circumstances, that the use of the process is prohibited;
- he offers, puts on the market or uses a product obtained by a process which is the subject matter of the patent, or imports or stocks the product for these purposes (indirect product protection; see Community Patent Convention Article 29c); or
- without permission, he exploits the invention by supplying or offering to supply any person who is not entitled to exploit the invention with means for working it in Denmark, if these means relate to an essential element of the invention and the person supplying or offering to supply the means knows, or it is obvious in the circumstances, that these means are suitable and intended for such use (ie 'indirect patent infringement', see Article 30 of the Community Patent Convention).

Acts which are carried out for non-commercial or experimental purposes, or concerning products put on the market in Denmark by the proprietor of the patent or with his consent, or in the preparation in a pharmacy of a medicine in accordance with a medical prescription, or acts concerning the medicine itself do not constitute infringement. The common defence arguments for infringers are invalidity and denial of infringement. A claim for invalidity can be based on the following (see Section 52 of the Danish Patents Act):

- the conditions for issuing a patent have not been fulfilled (ie in relation to novelty and inventive step;

- the patent is not described clearly enough for an expert to carry out the invention on the basis of the description;
- the object of the patent exceeds the contents of the application filed; and
- the scope of the patent protection has been expanded (due to a mistake) after grant of the patent.

Both the patentee and a third party can ask for an administrative re-examination of a patent already granted. The Patent Office may deem the patent invalid or uphold it in its original, modified or revised form.

When deciding whether a third party is infringing a patent, the patented claims are compared with the potentially infringing product or process. The scope of protection is determined by the patent claim with guidance from the description (see Section 2 of the Patents Act and also the Community Patent Convention, Art 69 with protocols including the EPC 2000 addition on equivalence).

The fundamental basis of the interpretation of patent claims is the natural linguistic meaning, but the description should be used as guidance for the interpretation. The Danish construction of the claims complies with the outline set out in Article 69 of the Community Patent Convention and is thus not literal or in accordance with the previously applied core theory, but it attempts to find the invention's real contribution to technical development. The courts will give wider protection to what are considered the essential elements of the patent, and infringement may be found even though unimportant features of the patent claim are not present in the allegedly infringing product or process etc.

A doctrine of equivalence applies in Danish patent litigation, and substitutions which may be perceived as technical equals do not remove the product or process from the protected scope of the patent. The doctrine of equivalence thus includes products that do not infringe the patent in a strict literal meaning. The doctrine of equivalence probably complies with the procedure set out in the protocol to Article 69 of the Community Patent Convention. The principle is now part of Danish law, together with the rest of the Community Patent Convention.

8.2 Assessment of patentability

The patentability of a patent is assessed in such a manner that patents are granted for inventions which can be used commercially and are new and inventive in relation to the state of the art at the time of filing the patent application. (Inventions do not have to have inventive step over hitherto undisclosed patent and utility model applications.) Discoveries, scientific theories and mathematical methods, aesthetic creations, schemes, rules or methods for performing mental acts, playing games or trading, and programs for computers or presentations of information are excluded as patentable inventions. Also excluded are methods for the treatment of the human or animal body by surgery or therapy and diagnostic methods practised on the human or animal body, although products, including substances and compositions, for use in any of these methods may be patented. Patents may not be granted in respect of inventions which would be contrary to public policy or morality. Neither are patents granted for plant or animal varieties or essentially biological processes for the production of plants or animals. Patents may, however, be granted for microbiological processes and the resulting products. With effect from July 30 2000, the Patents Act in Denmark implements the EC Directive on the protection of biotechnological inventions excluding patentability of the human body, its origin and development (including sequences of genes), except under certain circumstances. Furthermore, the Patents Act now specifically bans the patenting of inventions the commercial exploitation of which would contravene common decency or public policy.

Patents are granted after application to the Danish Patent and Trademark Office, to the European Patent Office designating Denmark or by means of an international patent application according to the PCT system (under the Patent Cooperation Treaty which was produced under the auspices of the World Intellectual Property Organisation – WIPO).

Novelty is assessed according to Section 2 of the Danish Patents Act, which requires objective global novelty. This entails that patents are granted if the invention is new compared with the state of the art before filing of the patent application, without limitations in regard to time or geographical condition. Denmark uses the principle of 'first to file' and not 'first to invent'.

There is inventive step if the invention has material distinctiveness over the state of the art. The criteria correspond to those set out in the EPO guidelines.

Denmark closely follows EPO practice concerning industrial applicability and practical use. The disclosure in the specification/description must enable a person skilled in the art to manufacture or carry out the invention. The question of adequate disclosure is assessed as of the date of filing. Sufficiency is decided by the court, judged through the eyes of the person skilled in the art. We are not aware of any decisions invalidating a patent for insufficient disclosure (enablement). If a patented invention does not have the required sufficiency, it can be declared invalid pursuant to article 52 of the Danish Patents Act. However, this requirement is considered to be of limited scope in Denmark.

9.   Available remedies

If a patent proprietor wants to counter infringements of his patent, he can make use of a range of remedies offered by Danish law. Courts in Denmark have powers to grant the following remedies: injunctions, delivery up and/or destruction of infringing items, pre-trial preservation of evidence, requested corrective measures, a declaration that intellectual property rights are valid and have been infringed, publication of court decisions, and damages. Typically in patent litigation, proceedings will contain at least a claim for injunction, a claim for damages, a claim for destruction of infringing items and a claim for recall.

An interim injunction is often the most important sanction for the patent proprietor since time is often of the essence and court-awarded damages in ordinary trials on the merits are often relatively modest (ie they are seldom considered to cover the full loss, although that is the theoretical staring point). An injunction is heard before the local Bailiff's Court.

The patentee may claim compensation from the infringer in the form of reasonable remuneration for the infringer's unauthorised exploitation of the patented invention and the additional damages caused by the infringement (Section 58 of the Danish Patent Act).

Access to remuneration has been part of Danish patent law since 1967. It ensures the patent proprietor a minimum award of an amount corresponding to reasonable royalties, even in cases where the patent proprietor has not suffered a provable loss. If the size of the reasonable royalty is not substantiated, the consideration is assessed according to the market value of the advantage obtained by the defendant by the infringement (see the Danish Supreme Court's decision reproduced in U 2007.1219 H (*Danish Weekly Law Journal* 2007, page 1219)).

Damages for further injury are awarded pursuant to the general law of tort. Following the amendment of the Danish Patents Act of January 1 2006 in connection with the implementation of the EU Enforcement Directive, 2004/48/EC, it is now explicitly stated in the law that the court can include the size of the infringer's unjustified profits as one of the parameters when awarding damages. This deviates from the general law of tort, which is based on a principle of redress. The new rule is expected to have an influence on the size of the damages awarded, since it provides the court with a legal basis to apply a more liberal assessment of the loss suffered. Furthermore, it will ease the onus of proof for the claimant.

Further, the injured party can be awarded remuneration for any non-financial damages ('moral prejudice') caused by the infringement in appropriate circumstances. To date, there have been no cases on this new aspect of the law, and it is unclear in what situation such damages may be awarded.

If the infringement is intentional, the patentee may also claim that the infringing party be fined. Under aggravating circumstances, the penalty may increase to imprisonment for up to one year for the person responsible for the infringement. The patentee cannot, however, initiate the process, but may request that the public prosecutor institutes such proceedings as a police prosecution.

The civil courts also have access to corrective measures when requested by the plaintiff. Corrective measures can only be applied to cases before the civil courts and not before the Bailiff's Court. Corrective measures in the Danish Patents Act correspond to Article 10 of the EU Enforcement Directive, but are not limited to just those measures. Under the directive, the courts must always make their decisions in accordance with the principle of proportionality. Additionally, the patentee may request the court to order the infringing party to publish the court's judicial decisions in accordance with Article 15 of the Enforcement Directive.

As a general rule, appeal against an interim injunction from the Bailiff's Court will not act as a stay of execution. However, the Bailiff's Court and the High Court are, in principle, not precluded from granting in an appeal a stay of execution.

Decisions pronounced by the civil courts must be fulfilled within four weeks from the District Court to the High Court and eight weeks from the High Court and the Maritime and Commercial Court to the Supreme Court. After the expiration of the term for appeal, decisions can be enforced. The court can rule that the decision is enforceable before the term of appeal has expired.

10.   Costs

A fee of Dkr500 is payable to the court. In cases where the value of the claim exceeds Dkr50,000, the fee payable to the court is Dkr750, plus an additional fee of 1.2% of the part of the claim exceeding Dkr50,000. However, the court fee cannot exceed Dkr75,000. Moreover, an additional court fee of Dkr750 must be paid when the date for the oral hearing of the case is fixed, plus a fee of 1.2% of the part of the claim exceeding Dkr50,000. Again, this fee cannot exceed Dkr75,000. For interim injunctions, a fee of only Dkr300 is payable to the Bailiff's Court. However, the same rules as for ordinary patent proceedings apply to the confirmatory suit, which has to be filed within two weeks of the grant to uphold the injunction.

The plaintiff must estimate the value of each claim separately at the time of submission of the writ. In relation to a claim for compensation, the value naturally forms part of this claim, while the value must be estimated in relation to declaratory claims and prohibition claims.

Each party shall pay its own costs incurred through its own actions. The costs are paid with the reservation that the party can demand compensation of its costs, if that party wins the case. The general rule is that the losing party is obliged to reimburse the other party's costs. However, in fact the court determines the costs awarded, and the amount awarded usually covers less than half of the actual costs payable to lawyers, patent attorneys and experts. Expenses not essential to the proper handling of the case are not reimbursed. The court decides which expenses were essential, including, for example, expenses relating to the attorney's fee. If the case is dismissed, the court can order one party to reimburse the other party's costs partly or wholly, or decide that each of the parties shall bear its own costs.

When a foreigner is the plaintiff to a case and the defendant makes a claim to this effect, the foreigner may at the discretion of the court be obliged to provide appropriate security for the defendant's costs which he may be ordered to pay. If Danish citizens in the foreign country in question are exempted from providing such security, the Danish courts cannot demand such security. If the security requested is not provided, the case is dismissed.

Both the plaintiff and the defendant can stop the case. If the defendant wishes to stop the case, he can admit the infringement or fail to appear before the court, in which case a judgment by default will be rendered. If the plaintiff wishes to stop the case, he can fail to appear and/or discontinue the action. In the latter case, the court will normally decide the cost issue in favour of the defendant.

In ordinary patent infringement proceedings, the plaintiff should normally expect to incur the further expenses relating to an expert appraisal, in addition to those fees payable to the court. The plaintiff is required to pay the experts' fee, but as a rule the unsuccessful party will in the end have to pay the costs of the court-appointed expert appraisal. Technical experts' fees in a reasonably straightforward mechanical case should be expected to be around Dkr150,000 to Dkr300,000. In patent infringement proceedings, the plaintiff's attorney will normally need to be assisted by a patent agent with the necessary technical expertise. The fee to the patent agent can be expected to be around Dkr200,000 to Dkr400,000 for one instance, or more in very complex matters. Under normal circumstances, the attorney's fees should be expected to be around Dkr500,000 to Dkr1,000,000, in first-instance proceedings.

In interim injunction cases, the defendant may be awarded legal costs if the injunction request is dismissed, but to date defendants have not been awarded costs relating to their own experts (this is subject to a decision of the Danish Supreme Court).

In ordinary patent proceeding cases, the successful party can sometimes claim costs for their patent attorneys' fees in relation to drafting questions in the expert appraisal; but generally expenses to a party's own patent attorneys and to experts consulted will not be covered.

11. Hot topic(s)

In Denmark, the question of whether a patentee holding a patent for a pharmaceutical product can prevent the marketing of a generic pharmaceutical is currently under consideration. The course of events leading to the sale of a pharmaceutical product is initiated by the filing of an application for a marketing authorisation. The marketing authorisation could be just an indication that Denmark is to be a reference member state for the generic product, or it could be the initial step on the way to marketing of the generic product in Denmark. The current question is how to substantiate the threat of a generic product coming onto the market before expiry of the patented product.

It is a fact that neither an application for, nor the grant of, a marketing authorisation in itself constitutes infringement and that it can therefore not form the basis of an interim injunction. However, both can be considered a threat to infringe as they indicate that the producer of the generic product is preparing to launch a product. Therefore, the application for, or grant of, a marketing authorisation may result in the patentee requesting assurance from the generic producer that the product in question will not be marketed before expiry of the rélevant patent/SPC rights. No reply, or an evasive reply, may constitute sufficient ground for establishing that there is an actual threat of infringement and might therefore form the basis for an interim injunction.

Two recent decisions have shed light on the legal situation in Denmark. In the first (*Novartis v Teva* on February 26 2008), an interim injunction was not granted owing to the Bailiff's Court's finding that it was of no immediate interest. The defendant had, at the time of the oral hearing, applied for but not been granted a marketing authorisation by the Danish Medicines Agency and had also informed the patentee that it had no current plans to market its product in Denmark (and that it did not infringe any valid patents), and that Denmark was primarily used as a reference member state in a decentralised application procedure.

In the second decision (*Eli Lilly v Nomeco* on April 28 2008), several generic producers of pharmaceutical products had filed applications for marketing authorisations for pharmaceutical products in Denmark containing the same active substance as that of the patentee. The patentee requested that the generic producers confirm that they would not commence marketing their products until the patent had expired. In their various replies they claimed that they were currently not planning to market a generic product and, if planning to do so, such marketing would not infringe any valid patents. The answers were not clear and unconditional, and reservations were made in respect of the validity.

The question was whether the grant of several marketing authorisations constituted such a threat that it had sufficient actuality, and thus that an unlawful act is not merely hypothetical. The defendant, one of the three Danish distributors, argued that the grant of a marketing authorisation does not constitute infringement and that if the producers should market a product at some future time, the decision should be rendered by the Bailiff's Court at that time.

The court ruled that, in order to grant an injunction against an infringement that has not yet taken place, the plaintiffs have to substantiate or show that it is probable that the defendant intends to commit an infringing act within a short period of time. Applying for and obtaining marketing authorisation does not constitute infringement of the plaintiffs' exclusive right. However, application for marketing authorisation constitutes a key act prior to marketing of pharmaceutical products in Denmark and creates a natural assumption that the applicant in question intends to commence marketing of the product claimed in the application. The generic product was in fact marketed in the United Kingdom by another generic company in spite of the plaintiffs' patent.

The court also found the existing statements from some of the producers that had applied for marketing authorisations to be conditional and thus only to invalidate the abovementioned assumption to some degree. Furthermore, more specific information was not available about the background for the applications for marketing authorisations in Denmark – for example, as a reference member state.

Accordingly, the court found that it had been rendered probable that there was a current risk that one or several of the companies having obtained marketing authorisations intended to commence marketing of a product containing the patented active substance, also taking into consideration the large number of applications for marketing authorisations, and therefore that there was a risk that marketing would take place within a short timeframe.

This threat formed the background for warning letters to the generic producers asking about their plans to launch generic products.

The content of warning letters may vary from a reference to the existence of the patent/patent application at risk of being infringed, to firm statements about alleged infringement and the consequences thereof. The common procedure is to send warning letters after having received information that an infringing marketing authorisation has been granted, or is pending. However, a marketing authorisation may often be obtained in Denmark to serve as a reference product without any intent to market in Denmark. The warning letter should only include correct statements and references to claims which are not expected to be invalid; but there are no requirements that the letter should include documentation that the patent is valid or analyses of alleged infringing products. However, it should be noted that unfounded allegations of infringement, made in bad faith, may be considered undue harassment and may be illegal under competition law and probably also under the Danish Marketing Practices Act.

The warning letter may include requests for further information about the alleged infringing products/methods and/or requests for relevant samples and materials. The court may later draw adverse inferences from a refusal to produce such information/samples/materials, to the advantage of the patentee.

The advantages of sending warning letters are that the alleged infringer may choose to cease the activity considered to be infringing, perhaps by changing its product or procedure, or by stopping marketing of the product/use of the method. The warning letter may also induce a settlement, including (eg a licence agreement).

The most important procedural advantage is that there is a basis for commencing an injunction procedure, or for avoiding the argument that a later interim injunction is time barred.

The main disadvantages of sending a warning letter are that the potential infringer may be induced to file an opposition against pending patent applications, challenge the validity of the patent, file a non-infringement suit, design around the patents invoked or add further distance to possible infringing elements, thus making it more difficult for the patentee to win a later case.

If the warning is not followed up by further action within reasonable time, the rights holder may risk not being able to enforce the patents by an interim injunction, as one of the conditions for an interim injunction in Denmark is that enforcement cannot await an ordinary trial. Passivity/acquiescence may also result in damages awards being reduced, or to forfeiture of the right to damages for the preceding period.

Denmark summary

- In Denmark, a patent litigation team normally comprises lawyers, patent attorneys and persons from the company responsible for the legal and/or technical side of the relevant patent(s).
- All patent trials consist of a preparatory phase and an oral hearing. A trial on the merits will usually take three to four years, depending on the preparatory phase, including the expert appraisal. It is, however, possible to apply a 'fast track' procedure at the Maritime and Commercial Court.
- The fastest and most commonly applied remedy is to request an interim injunction, which can be obtained within four to eight months.
- In the Danish Court system, interim injunctions are conducted before the Bailiff's Court, while matters on the merit are conducted before the Maritime and Commercial Court. In the Maritime and Commercial Court, a patent case is always heard by at least one legal judge and two expert lay judges.
- In cases on the merits, the court will appoint one or more independent experts to prepare a report answering written questions posed by the parties. The expert(s) will be asked to verify their report orally in court. In interim injunctions, each party is allowed to produce declarations from their own experts, and such declarations will normally be closely coordinated with a patent attorney.
- Danish patent law includes a doctrine of equivalence. The interpretation of patent claims under Danish patent law is considered consistent with EPC

Article 69, including the protocol. Patentability is assessed in such a manner that patents are granted for inventions having novelty, inventive step and sufficiency of disclosure in accordance with EPO practice.

- Patent litigation claims will often contain claims for an injunction, damages, destruction of infringing items and a claim for recall. In general, damages awarded in Danish patent decisions tend to be low, making the injunction the most frequently applied remedy. The EU Enforcement Directive is fully implemented in Danish patent law.
- In patent litigation it is possible to apply the rules on preservation of evidence. These rules are generally used where the patent proprietor has demonstrated that it is probable that the defendant has infringed the rights in question and there is reason to believe that evidence of the alleged infringement and its possible extent can be found at the stated premises. Such proceedings have to be followed by a confirmatory action in order to be able to use the evidence later at trial.
- In patent litigation it is also possible to apply the rules on disclosure, which allow the court to order the defendant or a third party to disclose information in regard to goods or services that constitute an infringement. However, the principle of proportionality requires the court to consider the overall balance of convenience of the parties.
- The costs of a patent trial will vary from approximately Dkr500,000 to Dkr2,000,000. In patent litigation, the costs of patent attorneys (who assist with the phrasing of the questions to court-appointed experts) will usually be considerable. An interim injunction, followed by a confirmatory action, can be obtained at less cost; but it will still depend much on the subject matter and complexity of the relevant case.

Finland

Rainer Hilli
Roschier, Attorneys Ltd

---

1. Lawyers

In Finland, patent litigators generally do not have a technical education and this means that the litigation team usually consists of lawyers (*asianajaja*) and patent agents (*patenttiasiamies*) in close cooperation with the client.

Finland does not impose restrictions on rights of audience and, therefore, lawyers, patent agents and others may appear before the court.

2. The court system

2.1 Competent court

Under article 65 of the Finnish Patents Act, the District Court of Helsinki is competent to hear proceedings in respect of:

- proper title to the invention for which a patent is sought;
- invalidation, or transfer of a patent;
- grant of compulsory licences, and determination of new conditions for, or revocation of, such licences;
- infringement;
- declaratory judgments; and/or
- assessment of compensation and damages.

In addition, the District Court of Helsinki is competent in matters concerning rights relating to an invention for which a European patent under the European Patent Convention (EPC) has been filed. A prerequisite for hearing such a case in the District Court of Helsinki is that the defendant is domiciled in Finland, or that the plaintiff is domiciled in Finland and the defendant is not domiciled in a country which is party to the European Patent Convention. The District Court of Helsinki also hears cases where the parties to the dispute have agreed that it is to be the competent court. However, such a dispute may not be heard by the District Court of Helsinki if the same dispute between the same parties is pending before the court of another country which is a party to the European Patent Convention. If the competence of such foreign court has been contested, the District Court of Helsinki must postpone the hearing of the case until the question of competence has been finally decided upon by the foreign court.

Generally, the final decision of a court of a contracting state of the European Patent Convention, in the forms of dispute referred to above, will be enforceable in Finland.

The District Court of Helsinki has appointed a Division of the Court (the Third Division) to handle patent and other intellectual property (IP) matters.

In patent proceedings, the District Court is assisted by two technical experts appointed by the court. These experts give their views on the matters submitted to them by the court. Their views are entered in the court records, but such experts are not judges. They are, however, entitled to question the parties and the witnesses.

Foreign decisions are frequently referred to in patent proceedings and are often regarded as relevant, particularly if they are from the Nordic countries.

2.2 Role of Patent Office in litigation

In invalidation proceedings (regarding both national and European patents designated for Finland), the court will request the opinion of the Patent Office with regard to the validity of a patent which is subject to any claim of invalidity. In practice, the opinion of the Patent Office is requested by the court before the main hearing.

In infringement proceedings, the court may request the opinion of the Patent Office at the request of one of the parties. The court usually follows the opinion of the Patent Office, but it is not obliged to do so.

2.3 Stay of proceedings

Under the Patents Act, the court can stay infringement proceedings if invalidation proceedings against the patent-in-suit are initiated as a defence. The District Court of Helsinki has announced that as from January 1 2008 it will, contrary to its previous practice, not stay infringement proceedings if invalidation proceedings are initiated against the patent in suit; the matters will, if possible and reasonable, be handled jointly. If opposition proceedings are filed or are pending against a national or European patent which is part of infringement proceedings, the District Court may also suspend the infringement proceedings; but, in light of recent practice, this is unlikely to happen. If national or European Patent Office (EPO) opposition proceedings have led to the patent-in-suit being declared invalid, the Finnish court would be likely to stay infringement proceedings until any appeal against such decision has been ruled upon.

2.4 Appeals

Decisions of the District Court of Helsinki can be appealed to the Helsinki Court of Appeals. Proceedings before the Court of Appeals are in writing; but if a party so requests, a full review with an oral hearing is generally conducted. The appeals procedure generally takes 18 months to two years. Preliminary injunction appeals are handled more speedily (ie six months to a year). The term for filing an appeal is 30 days after the date of the ruling, provided a notice of appeal has been filed with the District Court within seven days from the ruling. The other side can then file a cross appeal.

2.5 Supreme Court

Rulings of the Court of Appeals can be appealed to the Supreme Court. However,

Supreme Court proceedings are subject to leave. Leave is granted if the matter is of a precedential nature. The Supreme Court thus decides on issues of law. The term for filing the application for leave and appeal is 60 days from the date of the ruling of the Court of Appeals.

2.6 Warning letters

The Finnish Bar Rules recommend that, before taking court action, the party initiating the action should notify the counterparty unless there are particular reasons (such as the urgency of the matter in hand) not to notify. When sending warning letters, it is important to draft them in such a style that they are not contrary to good business practice. In a recent ruling, the Supreme Court ruled that a warning letter containing unjustified and threatening language was contrary to fair business practice.

3. Procedure and timescale of proceedings

3.1 Main proceedings

Substantive infringement proceedings, as well as declaratory or revocation actions, start with the filing of an application for summons with the District Court of Helsinki. The application for summons should be drafted to include details of claims, an estimate of the amount of compensation claimed and relevant evidence. The District Court is responsible for the service of the summons on the defendant and allows time (30 to 60 days) for a written response. There may be a further exchange of briefs, after which the District Court will call the parties for a preparatory oral hearing. At this hearing, the parties present their claims, briefly state grounds for their claims and disclose what evidence, witnesses and/or experts they will rely upon in the proceedings. The date(s) for the main oral hearing is also set at this hearing.

At the main oral hearing, the parties present their claims, argue their case and present their witnesses. Ordinary patent proceedings generally take 18 months to two years before the District Court of Helsinki, though they do occasionally take longer. Accelerated proceedings are not available.

Appeals proceedings generally take 18 months to two years before the Court of Appeals. The Supreme Court handles an application for leave in three to six months and, if appeal is granted, a judgment is usually rendered within about a year.

4. Availability of pre-action evidence gathering/disclosure

It is possible for the patentee to secure evidence in pre-trial proceedings. The securing of evidence can be applied for before the relevant district court at the location where the evidence to be secured is located. A precondition for obtaining an order securing evidence is that evidence is likely to be destroyed and that infringement is likely. In practice, this makes it difficult to obtain documentary evidence prior to the main action. If an order to secure evidence can be obtained, such evidence can be declared confidential by the court if the evidence is considered to be a trade secret. If there is no secrecy order, it is possible to export secured evidence to foreign jurisdictions.

5.   Availability of interim relief

Interim injunctions can be granted by virtue of the general precautionary measure in Chapter 7 Section 3 of the Procedural Code. Three requirements must be fulfilled before an interim injunction can be granted under this provision:

- there must be a claim,
- the patentee must be able to show that there is a danger that his IP rights are adversely affected; and
- the court must balance the interests of the parties.

Under the first of the above requirements, the proprietor of the patent in question must establish that he has a registered patent right and that it is likely that the patent is being infringed by the other party. The second requirement is that there must be a danger that the opposing party undermines the value of the patentee's IP rights (by deed or action, or through negligence, or in some other manner), or reduces their value or significance. The patentee does not have to prove that the danger actually exists, but rather that there is a potential danger and that it is imminent.

Under the third requirement under Chapter 7 Section 3 of the Procedural Code, the court must be satisfied that the other party to the proceedings does not suffer undue inconvenience in comparison with the benefit to be secured. This means that the court must consider the interests of both parties. A prerequisite for granting an interim injunction is therefore that the infringement must be deemed to cause significant harm to the rights holder in comparison with the harm likely to be caused by the injunction to the alleged infringer.

Interim injunction proceedings generally take about three to six months, but *ex parte* injunctions may be granted more speedily (ie in a couple of days). However, the courts tend to give the accused party the right to be heard in interim injunction proceedings, which relate to patent infringement. An *ex parte* injunction can be granted on a temporary basis without hearing the opposing party, in situations where the purpose of the injunction may otherwise be compromised. On the other hand, the patentee is strictly liable for damages if it subsequently turns out that the interim injunction was unfounded.

It is possible to obtain an interim injunction before any actual infringement takes place, if the patentee can prove that there is a threat of patent infringement. The limited court practice in Finland on precautionary measures in patent infringement matters indicates that there is a high standard of proof as regards both the existence of a threat of infringement and the potential harm likely to be suffered by the patentee as a result of such infringement. Therefore, although in theory it may be possible to obtain an interim injunction (eg at the stage before import and/or marketing commences), in practice it is unlikely that courts would grant such an injunction unless it is probable that the product in question will infringe relevant IP rights and that the adverse party intends to commence import and/or marketing in Finland prior to the expiry of the patent.

In its only decision regarding interim injunctions based on patent infringement, the Supreme Court of Finland has set the threshold for establishing the applicant's probable right at a high level. As a consequence of the ruling, the courts in Finland have become hesitant to grant interim injunctions in patent infringement matters, especially as regards infringement of process patents. The Helsinki Court of Appeals has, however, subsequently granted a preliminary injunction on the basis that an infringement was more likely than not (ie more than a 50% likelihood). At present, it is doubtful that Finland is fulfilling the obligation imposed on member states by the TRIPS Agreement (on Trade-Related Aspects of Intellectual Property Rights) and the EU Enforcement Directive to ensure that rights holders are granted fast and efficient precautionary measures.

Protective briefs have been filed, but as a rule they are not effective before the Finnish courts as there is no recordal system before an actual case is pending.

Relief from a decision regarding an injunction or seizure, or the lifting of such measures during the course of the proceedings, may be sought by separate appeal.

6. Disclosure

Finnish courts observe a general principle of evaluation of evidence. Thus, in infringement proceedings the claimant must provide evidence to show that his IP rights have been infringed; and in patent invalidity proceedings the claimant must show a likelihood that a patent is not valid.

If a patent has been granted for a process for obtaining a product, any identical product produced without the consent of the proprietor of the patent shall, in the absence of proof to the contrary, be deemed to have been obtained by the patented process. In the submission of proof to the contrary, the legitimate interests of defendants in protecting their manufacturing and business secrets shall be taken into account.

It is possible during main proceedings to request disclosure of specific documents and to request on-site inspections. Such request is granted if properly specified, and if the information requested to be disclosed is relevant as evidence in the matter. Such documents or other evidence obtained can be used in proceedings in other jurisdictions, unless such documents or other evidence has been declared confidential.

7. Evidence

Preliminary injunction proceedings are, as a rule, documentary. This means that written evidence and statements by witnesses and expert witnesses are allowed.

In the main proceedings, the parties put both written and oral evidence before the court. Witnesses and expert witnesses are subject to cross-examination. Court-appointed experts may also ask questions of the party-appointed experts.

8. Law

Patent legislation in Finland is governed by the Finnish Patents Act (the "Patents Act") (550/67) of December 15 1967 as amended, and the Finnish Act on Employee Inventions (656/67) of December 29 1967 as amended. Finnish patent legislation is applicable in the Republic of Finland.

8.1 Validity

Anyone who has conceived an invention which is susceptible to industrial application is entitled to the exclusive right to exploit the invention commercially.

According to article 2 of the Finnish Patents Act, patents may only be granted for inventions which are new in relation to what was known before the priority date[1] of the patent application and which also differ essentially from what was previously known. Finnish courts are not bound by the European Patent Office Guidelines or decisions of its Board of Appeals, but they generally take these sources into account when rendering decisions on novelty and inventive step.

Validity is frequently challenged in patent infringement proceedings. Article 61 of the Patents Act stipulates that where the defendant in infringement proceedings claims that the patent is invalid, the court may, at the defendant's request, stop proceedings until validity has been determined. In practice, courts have tended to stay infringement proceedings if the defendant has so requested, but with effect from January 1 2008 this practice was changed by the District Court (see Section 2.3 above). The invalidity action is filed before the same court as the infringement proceedings.

8.2 Partial invalidity

The Supreme Court has ruled[2] that a patent can be declared partially invalid in invalidity proceedings. Therefore, in exceptional cases, where the amendment of the claim is simple and only limits the scope of protection such partial invalidation is possible.

The entry into force of the Act Revising the Convention on the Grant of European Patents (EPC 2000), has resulted in increased possibilities for post-grant amendments in Finland. An amendment of the Patents Act came into force at the same time as EPC 2000 (ie on December 13 2007). This reform contains a limitation procedure similar to that which is set out in the EPC 2000. With effect from December 13 2007, it has been possible for the patentee to make post-grant amendment requests directly to the National Board of Patents and Registration (ie the local Patent Office in Finland) or, if an invalidation proceeding before the District Court of Helsinki has started, before the District Court. This procedure will not be available for third parties. Post-grant amendments of patent claims are effective retroactively (*ex tunc*).[3]

The new provision applies both to European and Finnish Patents (ie all patents in force in Finland). There is no court practice yet, but according to the preparatory works of the reform of the Patent Act (Government Bill HE 92/2005), a request to limit patent claims may only be made once during the invalidation proceedings of the patent in question, unless the plaintiff has presented new grounds for invalidation after the limitation request. Moreover, the limitation request may no longer be presented in the appeal proceedings. Since the scope of protection of patent claims is an essential factor when assessing points of dispute, the limitation request should be made during the preparatory stage of the proceedings (ie before the main hearing commences).

---

1   On February 19 1988, the Supreme Court (KKO: 1998:16) declared a patent invalid due to the fact that the proprietor sold 45 items prior to the priority date, without ensuring that the purchaser would keep the invention confidential.
2   Supreme Court June 18 1984 (KKO: 1984-11-117).
3   According to the present Finnish law a final judgment of a court regarding restriction or expiry of a patent is valid retroactively from the date of the patent application. This means that amendments are effective *ex tunc*.

As with limitation proceedings, in administrative proceedings before the Patent Office the amendment of patent claims will always result in a limitation to the patent claims. The invention disclosed in the amended patent should be so clearly described that a person skilled in the art would be able to use the invention based on the information given in the patent. The patent may not be amended so that it would include something that has not been disclosed in the patent application when it was filed. In addition, the amended claims should not extend the patent protection beyond the scope of protection of the granted, limited, or (in opposition proceedings) amended patent. Should the patent protection be extended in this way, this would qualify as a ground for revocation under the Patent Act.

If the court accepts a request to limit patent claims, the limited patent would be the basis for the further handling of the invalidation proceedings. If the request is not accepted, it will not be possible to appeal such decision separately, but the patent in force is the basis for the further handling of the matter. The parties may appeal the decision concerning the amendment of patent claims only after the decision as to invalidity has been given.

As from December 13 2007, it has also been possible for the patentee to present a request for the limitation of patent claims to the Patent Office in Finland, and have the limitation question handled in separate administrative proceedings. The purpose of the new provisions of the Finnish Patent Act is to harmonise the (administrative) limitation procedure in Finland with that of the European Patent Office.

According to the revised provisions of the Patent Act, the patent holder may submit a written limitation request together with amended patent claims to the Finnish Patent Office. The patentee does not need to present any specific reasons for doing so. A limitation request may also be made based on a European patent that is in force in Finland. The local Patent Office cannot consider a limitation request if opposition proceedings are pending before the local Patent office or the EPO. Furthermore, the request will not be considered in the Patent Office if a cancellation action is pending in court proceedings.

As with the limitation in court proceedings, the amended patent must be so clearly presented that a person skilled in the art would be able to use the invention based on the information disclosed in the patent. The patent may not be amended to cover something that was not disclosed in the patent application when it was filed. Furthermore, the amended claims cannot extend the scope of patent protection beyond the protection of the granted patent, or the scope of a patent amended in opposition proceedings. According to the revised Patent Act, it will be possible to present new limitation requests.

The Patent Office will not conduct new novelty and patentability examinations in connection with the limitation proceedings. However, the limitation of the patent must be real, and the patentee cannot reformulate the patent claims (eg by clarifying the claims or by using other terminology). However, the absence of novelty and patentability examinations may not result in the patent claims being extended in the limitation proceedings, and the extension of patent claims beyond the original scope would be a ground for revoking the patent.

The final decision of the Patent Office can be appealed, if the decision goes against the applicant. According to the revised Patent Act, post-grant amendments of patent claims are applied retroactively (*ex tunc*).

8.3 Infringement

The statutory provisions on direct and indirect infringement are set out in articles 3, 3a and 3b of the Finnish Patents Act. The acts that are listed in these provisions as constituting direct or indirect infringement are considered to be representative, but it is not an exhaustive list.

Under article 3, the exclusive right conferred by a patent results in an inference (with some exceptions) that no one may exploit an invention, without the proprietor's consent, by:

- making, offering, putting on the market or using a product protected by the patent, or importing or possessing such product for these purposes;
- using a process protected by the patent or offering such process for use in Finland, if he knows or if it is evident from the circumstances that the use of the process is prohibited without the consent of the proprietor of the patent; or
- offering, putting on the market, or using a product obtained by a process protected by the patent, or importing or possessing such product for these purposes.

The Supreme Court ruled in case KKO: 2003:127 that the fact that fire extinguishing equipment was ready for use (but had not been used) constituted an infringement.

The exclusive right does not apply to:

- use which is not commercial;
- use of a patented product that has been put on the market within the European Economic Area by the proprietor of the patent or with his consent (exhaustion);
- use in experiments relating to the invention as such;
- use in experiments and tests relating to an invention concerning a medicinal product required for an application for a marketing authorisation for a medicinal product; or
- preparation in a pharmacy of a medicine prescribed by a physician in individual cases or treatment given with the aid of a medicine so prepared.

Under the Patents Act, the scope of protection does not extend to experiments relating to the invention as such. However, there is to date no Finnish case law providing guidance to the exemption.

The so-called 'Bolar' provision set forth in the Medicines Directive (2004/EC) was implemented into the Patents Act on May 1 2006. This provision, as implemented in Finland, stipulates that necessary studies and trials made to fulfil the regulatory requirements for obtaining a marketing authorisation do not constitute infringement.

8.4 Scope of protection

According to article 39 of the Patents Act, the scope of protection conferred by a patent is to be determined by patent claims. The description may serve as guidance for interpreting the claims. There have been discussions in Finnish legal literature on whether the interpretation should be based on article 69 EPC and the Protocol on the interpretation of article 69 EPC. Case law does not provide a clear answer. However, preparatory work for a recent amendment of the Patents Act confirms that the interpretation should be in line with article 65 EPC.[4]

In infringement proceedings, a Finnish court will use the patent as granted as a starting point. If and when the patent claims need interpretation, the court will look at the description and drawings for further reference. Finnish courts do not interpret claims strictly literally and they tend to interpret them quite broadly.

The prosecution files play a role in determining the scope of protection of a patent. However, there is no case law that would indicate that a file wrapper estoppel would apply.

When considering the scope of protection, it should be noted that the judge or judges in infringement proceedings are assisted by two court-appointed technical experts. The technical experts render their opinion on the scope of protection as a basis for the court ruling made by the judge(s).

It is also possible to request the Patent Office to render its opinion in infringement proceedings (eg if questions concerning prior art and prosecution issues arise).

The doctrine of equivalence has been discussed in Finnish legal literature. The courts have been reluctant to apply this doctrine. However, in the preparatory work for the amendment of the Patents Act (Government Bill),[5] it has been stated that the scope of protection should include infringement by equivalent means. Such preparatory work is often used by Finnish courts in the determination of questions which have not been expressly mentioned in legislation, with a view to ascertaining the intent of the legislator.

8.5 Indirect infringement

A patent confers on its proprietor the exclusive right to prevent any person not having his consent from supplying or offering to supply any person not entitled to exploit the invention with the means of working the invention in Finland in relation to an essential element of the invention, where such other person knows, or where it is evident from the circumstances, that the means are suitable and intended for working the invention (ie indirect infringement). This provision does not apply where the means are staple commercial products, except where such other person attempts to induce the receiver to commit any of the acts referred to above. The provisions on indirect infringement have rarely been subject to litigation and the scope of indirect infringement therefore remains to be tested through court precedents.

8.6 Experiments

A party can rely on experimental work as evidence of infringement or validity.

---

4   Government Bill.
5   Government Bill.

Experiments undertaken by third parties are regarded as carrying more weight than experiments made by a party.

8.7 Compulsory licences

Chapter 6 of the Patents Act sets out provisions on compulsory licences. In addition to the specific provisions of the Patents Act, article 82 of the EC Treaty may in extraordinary situations be relied upon by a defendant if a patentee abuses a dominant position and refuses to grant a licence. Compulsory licences are rare in Finnish court practice. Provisions are as set out next.

(a) Non-use

Where three years have elapsed since the grant of the patent and four years have elapsed from the filing of the application, and if the invention is not worked or brought into use to a reasonable extent, any person who wishes to work the invention in Finland may obtain a compulsory licence to do so unless legitimate grounds for failing to work the invention may be shown.

(b) Dependent patents

The proprietor of a patent for an invention whose exploitation is dependent on a patent held by another person may obtain a compulsory licence to exploit the invention protected by such patent if deemed reasonable in view of the importance of the first-mentioned invention or for other special reasons. The proprietor of a patent in respect of which a compulsory licence is granted may obtain a compulsory licence to exploit the dependent invention unless there are special reasons to the contrary.

(c) Public interest

In the event of considerable public interest, a person who wishes commercially to exploit an invention for which another person holds a patent may obtain a compulsory licence to do so.

(d) Prior exploitation

Any person who was commercially exploiting an invention in Finland which is the subject of a patent application at the time the application documents were made available to the public shall (if the application results in a patent) be entitled to a compulsory licence for such exploitation, provided there are special reasons for this and also provided that he had no knowledge of the application and could not reasonably have obtained such knowledge. Such a right also extends, under corresponding conditions, to any person who has made substantial preparations for commercial exploitation of the invention in Finland. Compulsory licences may also relate to the period of time preceding the grant of the patent.

A compulsory licence may, as a general rule, only be granted to a person deemed to be in a position to exploit the invention in an acceptable manner and in accordance with the terms of the licence. Before filing a claim for a compulsory licence, a verifiable effort should have been made to obtain a licence to the patented invention on reasonable commercial terms. A compulsory licence does not prevent the proprietor of the patent from exploiting the invention himself, or from granting licences under the patent. A compulsory licence may only be transferred to a third party when it is transferred together with the business in which it is exploited or was intended to be exploited.

9. Available remedies

Finland has implemented EU Directive 2004/48/EC, the Enforcement Directive, but the implementation did not require any major amendments to the Patents Act.

9.1 Injunction

Under Section 57 of the Patents Act, the court may forbid any person who infringes the exclusive right afforded by a patent (patent infringement) from continuing or repeating the act. If an infringement is repeated or continued after the ruling, such infringement is likely to be regarded as intentional and punishable, under the Penal Code, by fines and imprisonment.

9.2 Compensation and damages

Any person who intentionally or negligently infringes a patent shall be liable to pay reasonable compensation for the exploitation of the invention and damages for other loss caused by the infringement. In cases where the negligence is slight, the compensation may be adjusted accordingly. Compensation for exploitation is usually calculated as a reasonable licence fee, based on the evidence given in the case.

A person found guilty of patent infringement that is neither intentional nor negligent shall pay compensation for the exploitation of the invention if, and to the extent that, this is deemed reasonable.

In Finland, damages are held at a reasonable level and may not include any punitive element. Due to a lack of court precedents, the courts do not have any consistent practice to which they can refer when considering damages awards. The claimant must, however, provide relevant evidence – for example, regarding royalty rates in the particular industry and losses caused by the infringement.

9.3 Seizure

At the request of the claimant, the court may order that the infringing objects be provisionally seized for the duration of the proceedings. Where the court so orders, the claimant is usually required to provide security for any damage or inconvenience that may be suffered by the other party as a result of the seizure.

Copies of decisions in patent matters are communicated by the court to the Patent Authority, with a statement as to whether the decision has become final.

10. Costs

In Finland, the loser pays the winner's costs. However, the courts can, and do, adjust such costs for various reasons.

The parties' costs for an ordinary patent infringement lawsuit before the District

Court are likely to be between €200,000 and €500,000; but in more complex matters the costs may exceed €500,000. Moreover, costs may easily double if appeals are made and a main hearing is also held before the Court of Appeals.

11. Hot topics

As Finland has been very late in approving pharmaceutical product patents, pharmaceutical products currently on the market are protected only by process patents. Finland accepted that product patents for pharmaceuticals could be applied for as of January 1 1995. Thus, the latest of these, if excluded by Supplementary Protection certificates, will expire during 2019. This means that generic pharmaceutical companies try to enter the market in Finland earlier than in countries with product patents for pharmaceuticals. As a result, patent litigation has increased and there has been discussion on whether Finland has implemented the TRIPS Agreement correctly.

France

Alexandra Neri
Herbert Smith LLP

1. Lawyers

Patent litigation teams in France are made up of *avocats* (attorneys at law) and patent agents where a case involves complex technical issues.

The parties do not have the right to defend themselves before courts handling patent litigation. Only *avocats* registered with one of the French Bars have a right of audience before these courts.

As regards first-instance proceedings, Law No 71-1130 dated December 31 1971 provides that an *avocat* has only a local jurisdiction to represent his client. Thus, an *avocat* appointed by a party for litigation that will take place outside his local jurisdiction will designate a second *avocat* in the relevant jurisdiction for the procedural aspects of the case.

Before the Court of Appeal, article 899 of the French Civil Procedure Code (CPC) provides that each party must be represented by an *avoué* in charge of procedural issues. Therefore, parties have to appoint both an *avoué* for the procedural aspects of the case and an *avocat* who will prepare the written arguments that will be submitted to the court and plead for the substantive aspects of the case.

A similar rule applies for proceedings before the *Cour de cassation* (which is located in Paris). Indeed, article 973 CPC provides that an *avocat aux conseils* (*avoués* and *avocats aux conseils* are separate) must be appointed. He is then the sole representative of the party before the court.

2. The court system

2.1 Overview of the civil court system

Seven *Tribunaux de Grande Instance* (Courts of First Instance) and their associated Courts of Appeal have exclusive jurisdiction with respect to all patent-related matters.

The seven *Tribunaux de Grande Instance* having jurisdiction are those of Paris, Lyon, Marseille, Bordeaux, Lille, Toulouse and Strasbourg. The seven relevant Courts of Appeal having jurisdiction are those of Paris, Lyon, Aix, Bordeaux, Douai, Toulouse and Colmar (article D 631-2 of the French Intellectual Property Code (CPI) and article D 211-6 of the French Courts Organisation Code deriving from Decree No 2008-624 dated June 27 2008).

These courts (First Instance and Appeal) are generally composed of three judges to rule on the merits. It is, however, possible that a reporting judge (ie a single judge) be in charge of ruling on the merits. Where it is planned that this is to occur, any party that believes that the case requires a full bench division may ask for it.

Regarding the *Cour de cassation*, Article L 131-6-1 of the French Courts Organisation Code provides that it is usually made up of five of its members.

Although the seven abovementioned Courts of First Instance and Courts of Appeal have exclusive jurisdiction with respect to patent litigation cases, it is not possible to consider each of them as a specialist court. Nevertheless, as 70 per cent of patent litigation cases are heard by the Paris and Lyon courts, these could be considered as specialist courts. For instance, the 3rd Chamber of the Paris Court of First Instance and the 4th Chamber of the Paris Court of Appeal only hear intellectual property (IP) cases. A high degree of competence is thus expected from the Paris and Lyon courts.

Even though a limited number of courts have jurisdiction for patent-related litigation, judges have no technical background as such. They can, however, order expert reports on technical issues. They are not bound by such reports and remain the sole decision makers of the court.

Appeals against first-instance decisions can be lodged before the competent Court of Appeal. It is up to the appealing party to decide whether to appeal in relation to the whole case or only part of the case. Appeals can be made on points of fact and on points of law.

An appeal must be lodged with the *Greffe* (court clerk) of the competent Court of Appeal within one month from the notification of the judgment to the parties' counsel (articles 528 and 538 CPC). This one-month period is extended by two months for parties living abroad (article 643 CPC).

Decisions of the Courts of Appeal can be appealed before the *Cour de cassation* within two months from the notification of the Court of Appeal's decision (article 612 CPC). The two-month extension set out in article 643 CPC also applies to appeals before the *Cour de cassation*. Although the Court of First Instance and the Court of Appeal decide on both facts and law, appeals to the *Court de cassation* can only be made on a point of law. If the appeal in law to the *Cour de cassation* is admissible, the court renders a reasoned decision to dismiss the appeal in law, or to set the case aside, either wholly or partly, and the case is then sent back to a Court of Appeal.

There is no such thing in France as a leapfrog procedure for patent litigation.

French case law merely provides principles for interpretation of the law. As a consequence, unlike common-law systems (applying the so-called *stare decisis* principle) a judge is not bound to follow a former decision. Indeed, articles 5 and 1351 CPC provide that each court decision merely has effect on the particular case in question. However, with respect to the mere interpretation of law, it is generally acknowledged that decisions rendered by the *Cour de cassation* are usually followed by inferior courts.

2.2 Patent-related actions

Patent-related proceedings for the most part comprise actions for patent infringement, for patent revocation, claims of patent ownership and for declaratory judgment of non-infringement.

In actions for patent infringement, the patent holder may file an action for patent infringement where the patent right is infringed. Although the principle is that licensees do not have the right to initiate patent infringement proceedings, a licensee will be entitled to act where cumulative conditions are fulfilled. These conditions are as follows:

- The licensee must have been granted an exclusive licence;
- The licence agreement shall not contain any stipulation preventing the licensee from initiating infringement proceedings; and
- The licensee may only initiate proceedings where the licensor does not initiate them after the licensee has sent him a formal notice to do so.

In addition, the licence must be registered with the French Intellectual Property Office (INPI), in order to be enforceable against third parties.

Non-exclusive licensees can intervene in proceedings initiated by the licensor in order to claim damages for losses they have suffered. (This takes the form of voluntary intervention by way of court submissions.)

In France, it is possible to bring an action for infringement on the basis of a pending application, but the proceedings are then automatically stayed until the patent has been granted.

Pursuant to article L 613-1 CPI, the exclusive right of exploitation takes effect as of the filing of the application. However, article L 615-4 CPI states that acts committed prior to the date on which the patent application has been made public or prior to the date of notification to any third party of a true copy of such application shall not be considered to prejudice the rights deriving from the patent.

Acts constituting patent infringement are defined by articles L 613-3 and L 613-4 CPI. These articles provide an exhaustive list of acts that are considered to be patent infringements.

Two main categories must be distinguished. The first relates to all acts infringing a patent that covers a product. Such acts are the manufacturing, offering, putting on the market or using the product protected by the patent, or importing or stocking a product for such purposes. The second category relates all acts infringing a patent that covers a process or method. Such acts are using the method, offering to use the method and offering, putting on the market or detention of the product obtained through the implementation of the method.

A third category exists relating to infringement by complicity (sometimes known as 'contributory infringement'). Under this category, it is forbidden to supply or offer to supply the means of implementing the invention.

Where the offer, commercialisation, use or detention of infringing goods is made by a person other than the manufacturer, such person shall only be liable if he has committed such acts in 'full knowledge' (ie such person had knowledge of the patent and of the infringing nature of the goods).

As regards revocation proceedings, any person having *lato sensu* an interest in such action may file an action for revocation (eg an alleged infringer as a counterclaim in an infringement proceeding, or a competitor who has received a cease-and-desist letter). Actions for revocation are often raised to provide a counterclaim to patent infringement actions.

Article L 613-25 CPI provides several grounds for revocation. Patents are subject to revocation where they fail to fulfil conditions of novelty, inventive step and susceptibility of industrial application.

A patent will also be subject to revocation where its description is not sufficiently detailed for a person skilled in the art to be able to carry it out, and where its subject matter extends beyond the content of the patent application as filed.

An action for claiming ownership of a patent is set out in article L 611-8 CPI. This allows the legitimate rights holder to claim ownership over a patent that has been unlawfully filed by a third person. The claimant has to prove that he is the legitimate rights owner. If the action succeeds, the right is transferred to the claimant, who recovers the property of the patent from its date of application. The statute of limitation for this action is three years from the publication of the granting of the patent. If the bad faith of the person who filed the patent application is proven, the statute of limitation runs as from the expiration of the patent.

It is possible to seek a declaratory judgment of non-infringement of a patent under article L 615-9 CPI. This article enables the person making or planning industrial use of a technique (manufacturing of a product or the carrying out of a method) in the European Community (person A) to file an action against a patent holder in order to demonstrate that his exploitation does not fall within the scope of its patent.

Case law (which is somewhat scarce on this action) attaches great importance to the proof to be brought in such claims. The proof of industrial use is evidenced by the effective manufacturing of a product or the effective implementation of a method. Proof of planned industrial use arises from serious and effective preparatory acts. Such acts are realised where the planned use is beyond the project step and the means for the industrial use are at least partially implemented.

This action is carried out in two distinct phases.

First, A informs the patent holder of the use he is making or planning. The patent holder then responds to A. Either he agrees that the use or planned use does not fall within the scope of his patent and such answer will have the same value as a settlement, or he answers that it falls within the scope of his patent.

Secondly, in the absence of a response from the patent holder, or if the patent holder has responded that the use or planned use falls within the scope of his patent, A may file an action before the relevant *Tribunal de Grande Instance* in order to obtain a judgment declaring that he is not infringing the patent.

This action can be filed either as a principal action, or as a secondary action in parallel to an action in revocation. This action does not prevent the patent holder from filing an action for patent infringement.

(a)   *Warning letters*

Warning letters are necessary in some cases. As mentioned above, in France, patent rights take effect as from the filing of the application (article L 613-1 CPI). However, pursuant to article L 615-4 CPI, acts committed prior to the date of notification to any third party of a true copy of such application shall not be considered as infringement. As a result, a patent application holder wishing to file a lawsuit on infringement grounds must send a warning letter to the alleged infringer in order to let him know of the existence of the patent application. The patent application holder will then only file a lawsuit as of this date if the alleged infringer does not cease to infringe the patent application.

Once the patent is granted, warning letters are not compulsory before filing a lawsuit. Indeed it might be tactically more effective not to send such a letter if the patent holder wishes to proceed to a seizure of infringing goods. In such cases, a warning letter would result in informing the presumed infringing party of the action that the patent holder intends to carry out. This would allow the alleged infringer to conceal proof of infringement.

Two other situations may involve sending a warning letter.

First, as discussed above, pursuant to article L 615-1 CPI:

"the offering for sale (…) of an infringing product, where such acts are committed by a person other than the manufacturer of the infringing product, shall only imply the liability of the person committing them if such acts have been committed in full knowledge of the facts."

In this case, sending a warning letter may be necessary in order to inform the assumed infringer of the patent rights in order to prevent him denying "full knowledge" in the event of litigation.

Secondly, as discussed above, article L 615-9 CPI provides that:

"… any person who proves industrial exploitation on the territory of a Member State of the European Economic Community, or real and effective preparations to that effect, may invite the owner of a patent to take position on whether his title could be invoked against such industrial exploitation, the description of which shall be communicated to him.

If such person disputes the reply that is given to him or if the owner of the patent has not taken position within a period of three months, he may bring the owner of the patent before the Court for a decision on whether the patent constitutes an obstacle to the industrial exploitation in question, without prejudice to any proceedings for the nullity of the patent or subsequent infringement proceedings if the industrial exploitation is not carried out in accordance with the conditions specified in the description referred to in the above paragraph."

An action for declaratory judgment of non-infringement therefore implies a prior "invitation of the patentee to take position". Note that this procedure is rare in France.

3. Procedure and timescale of proceedings

3.1 Typical timetable for a patent action

Initiating patent litigation implies first that a writ of summons and complaint (*assignation*) be served upon the defendant through a bailiff. Then a second original version of this *assignation* has to be filed before the competent *Tribunal de Grande Instance* in order to register the case in the court's case list.

This formal written statement must contain:

- the names, occupations, and addresses of the parties;

- the court before which the action is brought;
- a statement of the complaint with a summary statement of the grounds (*moyens*); and
- specific reference to the fact that the case may be decided on the sole basis of the opposing party's arguments, should the party fail to appear before the court (if the party is defaulting, the judge can rely on the arguments developed by the claimant).

A list of evidence and the actual evidence supporting the complaint must also be attached.

When there are several chambers, the *Greffe* (court clerk) will assign the case to the competent chamber of the court. A pre-trial judge (*Juge de la mise en état*) will be designated to hold procedural hearings.

It generally takes between six weeks and three months for the first procedural hearing to occur, during which time the judge will make sure that the defendant has appointed an attorney and that the claimant has communicated the evidence to the defendant.

The procedure is indeed divided into two phases:

- a pre-trial phase, which is a preparatory phase in which parties successively exchange *conclusions* (legal memoranda containing allegations of the facts and legal arguments) and pieces of evidence under the supervision of the *Juge de la mise en état*; and
- a judgment phase, in which oral pleadings take place before the decision on the merits is rendered.

The *Juge de la mise en état* will control case management and ensure that there are few opportunities for the parties to delay proceedings. In this respect, it should be stressed that in Paris, a proceedings calendar has been agreed upon by the Paris Court of First Instance and the Paris Bar.

The parties must observe strict deadlines to provide their *conclusions* and communicate their pieces of evidence. If not, the defaulting party will be subject to possible sanctions, such as the early closing of the procedural phase.

Generally, there is about a six-week period between the two procedural hearings.

According to new procedural guidelines as set by the Paris Court of First Instance, the claimant can only provide the *assignation* and one set of *conclusions* and the defendant two sets of *conclusions*. This new procedural rule implies that the *assignation* already involves all grounds, as there will only be one remaining set of *conclusions*. Even though this is a recent rule, it should be stressed that it is strictly applied by this Chamber, which will reject any demand for a third set of *conclusions*.

When the case is set and ready for trial, the *Juge de la mise en état* renders an Order closing the debates, and a date for the oral pleadings on the merits is then fixed. The parties cannot exchange further *conclusions* and pieces of evidence after closure of the debates. If such exchange happens after this Order, the exchanged documents will be rejected by the judge.

Depending on the level of complexity in the case, judges will usually give their decisions between one to three months after the hearing for oral pleadings on the merits.

Typically, first-instance proceedings run for about 18 months to two years, depending on whether or not the patent is technically complex. Appeal proceedings usually take approximately the same length of time. Proceedings before the *Cour de cassation* generally take about two years.

3.2 Availability of accelerated proceedings on the merits

Proceedings on the merits may be accelerated by way of an *assignation à jour fixe* (fixed-day summons). The claimant has to present an *ex parte* request to the court in order to be authorised to serve a summons for a specific date in the near future. This requires that the specific urgency of the matter be invoked in a written complaint filed with its supporting pieces of evidence.

3.3 Potential effect of any EPO opposition proceedings

According to article L 614-15 CPI:

> *"the Court hearing proceedings for infringement of a French patent which covers the same invention as a European patent applied for by the same inventor or granted to him or to his successor in title with the same priority shall stay proceedings until the date on which the French patent ceases to have effect in accordance with Article L 614-13 [ie either the date on which the period during which opposition may be filed against the European patent expires without the opposition having been filed, or the date on which the opposition proceedings are closed and the European patent maintained] or until the date on which the European patent application is refused, withdrawn or considered to have been withdrawn, or the European patent is revoked."*

In brief, the court hearing the infringement action would stay proceedings until the end of the opposition proceedings. However, according to case law, where there is no French patent which covers the same invention as the European patent at issue, the abovementioned article does not apply and the stay of proceedings is not compulsory. Nevertheless, when the European patent is the subject of opposition proceedings before the EPO, a stay of proceedings is generally (but not systematically) granted for the sake of good administration of justice until the end of the opposition proceedings.

4. Availability of pre-action evidence gathering/disclosure

An 'infringement seizure' (*Saisie contrefaçon*) is a very efficient method for garnering evidence prior to litigation.

Law No 2007-1544, dated October 29 2007, implementing Directive 2004/48/EC (the [EU] Enforcement Directive) amended article L 615-5 CPI on infringement seizure. The infringement seizure has to be allowed on petition by the president of the Court of First Instance having jurisdiction over the patent infringement. This petition must be filed by the holder of the patent or patent application.

Under the procedure, a bailiff will be appointed by the patent holder to carry out either a seizure of samples and a full description of the products, or a seizure of the available stock. The new provisions allow the seizure of all devices and tools used to produce or distribute the infringing products, and accounting documents.

Decree no 2008-624, dated June 27 2008, determines the period of time in which the seizing party must file an action on the merits in order to validate the seizure. Under article R 615-3 CPI (under the decree), this period is of 20 working days, or 31 calendar days if longer, from the date of the seizure.

If an action on the merits is not filed in due time, the seizure is null and void and the seizing party may be liable for damages.

Two points should be emphasised:

- First, it is important to note that the patent holder can appoint his own experts to accompany the bailiff during the seizure. This is important, as it has given rise to numerous court decisions under former Article L 615-5 CPI. The law is now clear on this point; and
- Second, French law does not mention that the party requesting the seizure has to present "reasonably available evidence to support his/her claims that his/her intellectual property right has been infringed or is about to be infringed" as mentioned in the Enforcement Directive. This decision of the French legislator seems appropriate. As the aim of the measure is to obtain proof of infringement, it could be considered as inconsistent to ask the patentee to provide such proof prior to the seizure.

The results of a valid infringement seizure can be exported to foreign jurisdictions.

5. Availability of interim relief (especially injunctions)

A preliminary injunction procedure is set out in article L 615-3 CPI (as amended by Law No 2007-1544).

Law No 2007-1544 implemented article 9 of the Enforcement Directive and substantially amended the then applicable preliminary injunction procedure.

According to two recent decisions (*Tribunal de Grande Instance de Paris, Evac Eurl v Jets Vacuum As*, January 22 2008, RG 08/50559 et 08/50571; *Cour d'appel de Paris, Schneider Electric Industries v Chint Europe Ltd*, December 21 2007, RG 07/09889), even though this law is immediately applicable as it does not need any implementing measure to be enforceable, it cannot be invoked in pending cases. Indeed, only mere procedural laws are applicable to pending cases, whereas Law no 2007-1544 provides new rules on the merits for the application of article L 615-3 CPI.

Pursuant to new article L 615-3 CPI, any individual with capacity to act on patent infringement grounds can file summary proceedings before the competent Court of First Instance in order to obtain a preliminary injunction. The court may order, under penalty of a fine, any measure against the alleged infringer aimed at preventing either an imminent patent infringement or the pursuit of asserted acts of infringement. The competent court can also order any urgent measures on petition without the defendant being heard, where circumstances require that these measures should not be discussed in open debate – in particular, where a delay may cause irreparable damage to the petitioner.

With regard to either of these actions, article L 615-3 specifies that the jurisdiction cannot order the measures requested unless evidence, which is reasonably accessible to the petitioner, shows that it is likely that his rights are being infringed or that such an infringement is imminent. The court may also grant provisional damages to the patent holder where the harm he has suffered cannot be seriously contested.

The new system is very advantageous for the patent holder.

The proceedings can be filed either before or after the action on the merits. Where the measures ordering the cessation of the unlawful activity are ordered before the introduction of an action on the merits, such actions on the merits must be introduced within 20 working days (or 31 calendar days if longer) as from the date of the court order. Otherwise, such measures can be declared null and void by the court, at the request of the defendant, without prejudice to any damages that might also be claimed (article R. 615-1 CPI stemming from Decree No 2008-624 of June 27 2008).

The defendant has the right to lodge an appeal and the court can make the enforcement of the injunction conditional on the posting of a bond by the patent holder. This bond can be used to compensate the defendant if the measures are eventually nullified or if the action on the merits is eventually rejected, provided the defendant has suffered some loss as a result of the measures.

To date, patent summary proceedings have been somewhat scarce in France and have rarely produced satisfactory results. The aim of the new law is to make this procedure more effective.

6. Disclosure

In France, there is no procedure akin to that of discovery proceedings in the United States or the United Kingdom. The parties are not obliged to disclose documents relevant to the case and can control the evidence they wish to use. They are also not obliged to refer to any facts or evidence during the litigation if it is not in their interests to do so. However, a party may be ordered by the court to disclose information. Pursuant to article 9 CPC, each party must provide evidence for the success of their claim.

Articles 15 and 132 CPC provide that the parties must exchange the documents proving their claims. Further, such exchange has to be carried out within a reasonable period of time prior to the hearing of the case. Indeed, because the French procedure is in writing, all arguments and documents must be discussed before the hearing of the case.

Each party can choose which documents it will provide and can therefore withhold evidence that may be detrimental to its own interests.

The French Civil Procedure Code thus provides means to obtain copies of any document which is in the possession of the other party, or held by a third party. Indeed, the judge can, on request and under the penalty of a fine, order a party to proceedings (articles 133 et seq CPC) or a third party (article 138 *et seq* CPC) to communicate any document of interest.

More specifically, it is worth noting that, under article L 615-5-2 CPI (deriving from Law No 2007-1544), if requested by the claimant, and provided there is no legitimate impediment, the competent court can order (under penalty of a fine if necessary) that information on the origin and distribution networks of the infringing products or methods be provided by the defendant and/or any other person who:

- was found in possession of the infringing products;
- was found to be using the infringing methods;
- was found to be providing services used in infringing activities; or
- was indicated as being involved in the production, manufacture or distribution of these products, use of these methods, or provision of these services.

Documents and information comprise:

- the names and addresses of the producers, manufacturers, distributors, suppliers and other previous holders of the products, methods or services, as well as the intended wholesalers and retailers; and
- information on the quantities produced, commercialised, delivered, received or ordered, as well as the price obtained for the products, methods or services in question.

7. Evidence

As stated above, all arguments must in writing. Therefore, evidence usually consists of documents and affidavits, and these must be communicated in due time in order to allow the other party to respond.

Experts' reports are rendered either by experts appointed by the judge or instructed by a party. It is important to emphasise that judges generally attach little importance to an expert report instructed by a party where such report is not corroborated by objective elements. Experts' reports may be ordered by the judge; but generally these are only in relation to the issue of what damages should be to allocated to the party whose IP rights have been infringed.

8. Law

8.1 Claim construction and doctrine of equivalence

Under French law, the general principle governing the construction of claims is set out in article L 613-2 CPI which reads as follows:

"The extent of the protection conferred by a patent shall be determined by the terms of the claims. Nevertheless, the description and drawings shall be used to interpret the claims."

According to French case law, interpretation of claims is limited to the substance of the claimed invention, and extreme rules of interpretation will be excluded. However, protection will not be limited to the literal meaning of the wording of the claims.

For instance, in its decision in *Dolle v Emsens* of October 11 1990, the Paris Court of Appeal declared that:

"Article 69 [of the EPC], as completed by its protocol, has chosen a middle way between a literal construction of the claim, in which the description and the drawings should be

used only to dissipate ambiguities, and a broad construction in which the claim would be used just as a guideline and in which the protection would extend to what, according to the skilled person, the patentee has intended to protect.

This compromise must ensure a fair protection for the patentee against the skill of the infringer to disguise infringement and enable third parties to know with certainty what is protected.

In view of Article 69, the Judge must construe the claims by reference to the description and to the drawings. He must give to the claim its full meaning, so that this condensed text is understood. The construction leads to define the substance of the claimed invention, without adding any element which the claim did not include and did not suggest."

Under French law, a claim is considered to cover not only all elements as expressed in the claim, but also equivalents. According to the French doctrine of equivalents, two means (*moyens*) are considered as equivalent, when despite having a different form or structure (*forme*), they perform the same function (*fonction*) in order to achieve a same or similar result (*résultat*).

The two different forms must fulfil the same function. In other words, both means must result in the same primary technical effect.

There is an infringement where the same function is fulfilled through two means having a different form or structure (provided the patent covers not only the form of the means, but also its function, and thus that the function of this means was not known in the prior art on the day the patent was filed).

For instance, there is an infringement when all the essential elements of the claims are identically reproduced, or produced via equivalents (*Cour de cassation*, Commercial Branch, September 24 2002).

8.2   Assessment of patentability

Patentability is assessed according to three distinct and cumulative conditions:
- novelty;
- inventive step; and
- susceptibility of industrial application.

These conditions have to be assessed by the judge in this specific order. Regarding novelty, an invention shall be assessed as novel where it is not part of the state of the art.

The state of the art is determined as:

"… everything made available to the public by means of a written or oral description, by use or in any other way, before the date of filing of the patent application. Additionally, the content of French patent applications and of European or international patent applications which designate France, which dates of filing are prior to the date referred to in the first sentence of this Article and which were published on or after that date, shall be considered as comprised in the state of the art." (Article L 611-11 CPI)

Novelty is thus an objective criterion that is fulfilled where the invention is not part of the state of the art at its date of application. The state of the art does not involve any territorial or time limit (looking backwards). It is all that has been made available to the public until the date of application.

Novelty is thus assessed by comparing the invention to already existing inventions. It must be emphasised that mere similarity between the invention and the state of the art is not sufficient to reject novelty. Neither are necessary combinations of existing inventions sufficient. Only full anticipation (*antériorité de toutes pièces*) will prevent an invention from being considered as novel. Full anticipation implies that the invention is entirely found in a single item of prior art. According to case law, the single item of prior art must thus:

- be certain;
- contain all the elements of the applied invention in the same shape; and
- seek the same technical result (Paris Court of Appeal, January 6 2006).

As regards inventive step, Article L 611-14 CPI provides that:
"... an invention shall be considered to involve an inventive step if, having regard to the state of the art, it is not obvious to a person skilled in the art."
This is the most difficult condition to assess. In contrast to the requirement of novelty, this condition implies a subjective assessment made by a person skilled in the art. The person skilled in the art is an averagely skilled man who has knowledge of the technical issue from his professional life. Moreover, this person is deemed to be able to use any knowledge acquired in his professional life. In effect, the invention will be obvious if this person is able to obtain the invention through his own knowledge.

As regards industrial application, Article L 611-15 CPI provides that:
"... an invention shall be considered susceptible of industrial application if it can be made or used in any kind of industry, including agriculture."
This last condition is easy to fulfil. It simply implies that the method or product can be employed by or through industry, which is obvious where the invention is of a technical character. This concept is very widely framed, as it covers any organised human activity. As a result, any product capable of being manufactured and any method having economic significance is susceptible of industrial application.

On the whole, it should be said that the INPI does not examine whether a patent application includes an inventive step. This condition is usually first assessed by the judge when the issue of the validity of a patent arises before him.

8.3 Patentability assessment of certain subject matter

As regards biotechnology, the new provisions derive from Directive 98/44/EC of the European Parliament and of the Council dated July 6 1998 "on the legal protection of biotechnological inventions" (hereinafter the EC Biotech Directive) and have been implemented in French Law by Law 2004-800 dated August 6 2004 "on Bioethics" and Law 2004-1338 dated December 8 2004 "on the legal protection of biotechnological inventions".

In light of paragraph 4 of the current version of article L 611-10 CPI, biological material is patentable if it complies with the requirements of:

- novelty;
- inventive step; and
- industrial applicability, subject to the provisions of Articles L 611-17, L 611-18 and L 611-19 CPI.

Article L 611-17 used to provide that the human body, its elements and products as well as knowledge of the whole or part of a human gene cannot, as such, be subject to patents, and that plants and animal varieties are not patentable. It no longer does so. Such exclusion from patentability is now provided by articles L 611-18 and L 611-19 CPI, as set out below.

Article L 611-17 CPI reads as follows:

"Inventions shall not be considered patentable where their commercial exploitation would be contrary to human dignity, public order or morality; however, exploitation shall not be deemed to be so merely because it is prohibited by Law or regulation."

According to Article L 611-18 CPI:

"The human body, at various stages of its formation and development, and the simple discovery of one of its elements, including the sequence or partial sequence of a gene, cannot constitute patentable inventions.

Only an invention constituting the technical application of a function of an element of the human body may be patentable. The protection will only cover the element of the human body to the extent of what is necessary to realisation and exploitation of this specific industrial application. This industrial application must be concretely and precisely exposed in the patent application.

Notably, the following shall not be patentable:

(a) processes for cloning human beings;

(b) processes for modifying the germ line genetic identity of human beings;

(c) uses of human embryos for industrial or commercial purposes;

(d) sequences or partial sequences of a gene, as such."

This article was introduced into the CPI by article 17 of Law 2004-800. It is more restrictive than the directive, which recognises the patentability of isolated elements and especially cells.

According to article L 611-19 CPI:

"I – The following shall not be patentable:

1° animal varieties;

2° plant varieties as defined by Article 5 of EC Council Regulation No 2100/94 of 2/ July 1994, instituting a system of community plant variety protection;

3° essentially biological processes for the production of plant varieties and animals; shall be considered as such processes those which consist exclusively of natural phenomena such as crossing or selection;

4° processes for modifying the genetic identity of animals which are likely to cause them suffering without any substantial medical benefit to man or animal, as well as animals resulting from such processes.

II – Notwithstanding provisions of I, inventions based on plants or animals are patentable if the technical feasibility of the invention is not limited to a specific plant or animal variety.

III – Provisions of 1–3° shall not prejudice the patentability of inventions which concern a microbiological process or other technical process or a product obtained by means of such a process; shall be regarded as microbiological any process involving or performed upon or resulting in biological material."

This article derives from article 17 of Law 2004-800, as amended by article 2 of Law 2004-1338.

9. Available remedies

The Enforcement Directive 2004/48/EC was implemented in France via Law 2007-1544. This law modified articles L 615-7 and L 615-7-1 CPI on available remedies.

Article L 615-7 CPI specifically deals with damages. The aim of damages is to repair both material loss and moral prejudice (eg damage to reputation).

In assessing damages, the judge will take into account both the negative economic consequences of the infringement (ie the lost profits of the patent holder and the unfair profits made by the infringer), and the moral prejudice resulting from the breach of the intellectual property right.

As an alternative, and where the plaintiff specifically asks for it, the calculation of damages may consist of the judge setting the damages as a lump sum equal to or greater than the amount of royalties or fees which would have been due if the infringer had requested authorisation to use the patent in question. Although case law implementing this provision is not yet available, former decisions were already based on such means of calculation. Indeed, some courts relied on the rate of an average licence to determine the economic prejudice. The courts applied an average rate to the amount of infringing products in order to determine lost profits.

Article L 615-7-1 CPI grants additional remedies. First, the judge may order that infringing products and the means to manufacture such products be recalled and withdrawn from commercial circuits, destroyed or confiscated to the benefit of the plaintiff. Secondly, the judge may order publication of the judgment in appropriate journals or newspapers. Such additional measures can only be decided by the judge where the plaintiff asks for them. The infringer must bear the corresponding costs.

9.1 Provisional enforcement

Under article 539 CPC, the enforcement of a decision on the merits is automatically suspended when an appeal is lodged against it.

However, article 524 CPC provides that a first-instance judgment is enforceable despite the appeal where provisional enforcement has been pronounced. Indeed, article 515 of the CPC allows that provisional enforcement "be ordered, at the request of the parties or *sua sponte*, each time the judge deems it proper and compatible with the nature of the matter".

10. Costs

First-instance infringement proceedings generally cost between €30,000 and €150,000 (including attorneys' fees and patent agents' fees), depending on the complexity of the case. Appeals will generally cost a further €30,000 to €80,000.

Attorneys' fees can be specified by way of an hourly rate, a lump sum or a lump sum with a contingency fee (but not on the basis of contingency fees only).

As a general rule, the unsuccessful party will have to bear the court costs (*dépens*), including fees for pleading in court, but not attorneys' fees unless the court orders otherwise. Under French law, each party must cover its own legal costs, including attorneys' fees. French civil procedure rules do, however, contain a provision (article 700 CPC) which states that the court may order the unsuccessful party to bear the legal costs of the successful party.

11. Hot topic(s)

After several years of discussion, the French Bar should soon merge with the IP attorneys' profession (ie patent agents and trademark and designs attorneys). The French Bar Association has stated that it will allow those who are already registered IP attorneys to join under certain conditions (including stipulations as to training); and on September 12 2008 it approved a report containing proposals for amendments to applicable legislation. On October 15 2008, the French Institute of IP Attorneys (CNCPI) voted on this report and approved it (ie the majority decided that IP attorneys should join the Bar). A Bill has been drafted and, at the time of writing, is currently being reviewed by the Ministry of Economy, Industry and Employment and the Ministry of Justice.

France summary

- Cases are heard by non-specialist/technical judges, but the Paris/Lyon courts have experience in patent litigation.
- Initiating patent litigation involves service of a writ of summons and complaint on the defendant, through a bailiff. The first procedural hearing will generally take place within three to six weeks.
- The process of *Saisies contrefaçon* is a powerful tool for evidence gathering.
- Preliminary injunctions have traditionally been hard to obtain, but the process is becoming easier.
- Depending on the complexity of the case, judges will usually render their decision from one to three months after the hearing of oral pleadings on the merits.
- Typically, first-instance proceedings will run for about 18 months to two years depending on whether or not the patent is technically complex. Appeal proceedings usually take approximately the same time. Proceedings before the *Cour de cassation* (Supreme Court) generally take about two years.
- First-instance proceedings generally cost between €30,000 and €150,000 (including attorneys' fees and patent agents' fees), depending on the complexity of the case. Appeals generally cost between €30,000 and €80,000.

Germany

Thomas Bopp
Henrik Holzapfel
Gleiss Lutz

1.   Split nullity/infringement system

A fundamental feature of the German patent system is that actions on the merits for patent infringement and proceedings for patent invalidity are heard separately from one another and before different courts/instances. In an infringement action on the merits, there is no invalidity defence. See Section 4 below.

2.   Lawyers

In German patent infringement litigation, each party must be represented by a lawyer (*Rechtsanwalt*). Usually, the team in patent litigation will also include a patent attorney (*Patentanwalt*) to support the lawyer. In a patent revocation action, each party may be represented by a lawyer and/or by a patent attorney.

Lawyers have rights of audience in every court, and there is no need to instruct 'local' lawyers. However, one exception to this applies to the final instance in patent infringement proceedings. In the case of proceedings before the Federal Court of Justice (*Bundesgerichtshof*), each party must be represented by a specialised lawyer. Other lawyers are not admitted to this court.

3.   The court system

3.1   Patent infringement cases

Patent infringement cases are heard in three instances: District Courts (*Landgerichte*), Higher Regional Courts (*Oberlandesgerichte*) and the Federal Court of Justice.

At first instance, there are 12 District Courts which are competent and specialised for patent infringement litigation (roughly one court per federal state (*Bundesland*)). The litigator has a choice of forum if, as is usually the case, the infringing acts occur nationwide. Three professional judges make up a District Court's chamber. There is no jury.

All judges are lawyer-judges. Judges are not simply senior lawyers; they are a separate profession with specialised training. In general, these lawyer-judges do not have a technical background. Nevertheless, some of the most frequently chosen District Courts such as Düsseldorf and Mannheim hear over 100 patent infringement cases per year, and they have developed a good understanding of technical issues.

The District Court of Düsseldorf, in particular, has an excellent reputation internationally for its expertise and the quality and predictability of its decisions. The German courts currently hear more patent infringement cases than those of any other European country.

From the District Courts, an appeal lies to the specialised senates of Higher Regional Courts (one per District Court – for example, from the District Court of Düsseldorf to the Higher Regional Court of Düsseldorf). In general, the appeal court is bound to the findings of fact of the court of first instance. Exceptions apply if the findings of fact are incomplete. However, a party may only introduce new facts in appeal proceedings if it has not acted negligently in not presenting the facts in first instance. Like the first-instance court, the appeal court sits with three professional lawyer-judges. Again, the judges are experienced in patent infringement cases.

From the Higher Regional Courts, a further appeal is possible to the Federal Court of Justice, which is located in Karlsruhe. Such further appeal is restricted to a legal review and is only admissible if:

- the matter is of general relevance; or
- the appeal is desirable in order to develop a point of law or to make the law more consistent; or
- the appeal is desirable in order to correct significant errors in the appeal judgment.

Such appeals in patent matters are heard by the 10th Senate of the Federal Court of Justice, which sits with five professional lawyer-judges, all of whom will have extensive experience in patent matters.

In theory, a leapfrog appeal that leads from the District Courts directly to the Federal Court of Justice is also possible. However, such leapfrog appeals are rarely practised, as they require consent from both parties and from the Federal Court of Justice.

Only one judgment of the court is given in patent infringement cases and there are no dissenting opinions of individual judges.

3.2 Patent revocation cases

Patent revocation cases are heard in two instances: the Federal Patent Court (*Bundespatentgericht*) and the Federal Court of Justice.

Only the Federal Patent Court, which is located in Munich, is competent at first instance. The Federal Patent Court has specialised senates for the different fields of technology. Five professional judges make up a senate. Two of them, including the presiding judge, are lawyer-judges, while the remaining three have a technical background.

From the Federal Patent Court, an appeal lies to the 10th Senate of the Federal Court of Justice. Such appeal is not restricted – at least under present law (as at September 2008) – to a legal review. The parties are free to introduce new facts on appeal.

Again, in patent revocation cases, only one judgment of the court is given. No dissenting opinions of individual judges are given.

3.3 Actions on the merits

Two types of actions on the merits are available in patent infringement litigation.

First, a patent proprietor or licensee under a patent can file a suit for patent infringement. In such a suit, the plaintiff typically claims for, *inter alia*:

- cease and desist;
- information and rendering of accounts; and
- a declaratory judgment on damages (see Section 10 below).

Secondly, an alleged patent infringer can file a suit for declaration of non-infringement. The condition for such declaratory judgment is that the plaintiff has a legal interest in the judgment, for example because it has received a warning letter from the patent proprietor. In response to a suit for declaration of non-infringement, the patent proprietor or licensee under a patent may file a suit for patent infringement. As soon as this suit has been tried in an oral hearing, the alleged infringer loses its legal interest in an independent declaration of non-infringement and consequently has to declare the suit for declaration of non-infringement terminated.

A patent revocation action can be filed by any party at any time, at least as long as the patent is in force.

3.4 Decisions from other jurisdictions

German courts usually take note of decisions from other jurisdictions, especially if the patent in suit is a European patent. Decisions of UK and Dutch courts are particularly persuasive for German courts. Decisions of US courts are taken into consideration to a lesser extent, as substantive US patent law differs considerably from both European and German patent law.

3.5 Requirements before filing an action

There are no special requirements that must be satisfied before an action can be filed. In particular, neither an attempt to reach an amicable solution nor a warning letter is required. Nor are they of particular practical importance. Sending a warning letter prior to commencing an action does have the advantage that the defendant would have to bear the costs of the proceedings, even if it immediately acknowledged the claims in court. However, such acknowledgement is rarely given. Furthermore, a warning letter has the disadvantage that it may give the infringer the opportunity to launch a so-called 'torpedo' (ie to file a suit for a declaratory judgment of non-infringement in another European country). If proceedings in the other country are slow, even an inadmissible suit for a declaratory judgment may considerably delay a patent infringement suit in Germany. This is a much-disputed effect of relevant European procedural law (ie Council Regulation (EC) No 44/2001).

4. Procedure and timescale of proceedings

In the case of both an action on the merits for patent infringement and a suit for declaration of non-infringement, a judgment at first instance typically takes approximately 8 to 12 months. The duration of the proceedings depends mainly on the chosen forum. If a neutral expert is appointed by the court, the duration may exceed this kind of timescale by up to one or two years.

Following the filing of the action, the court will serve the statement of claims upon the defendant within about one week. If the defendant is located in another European country, service will take approximately one or two months. The timetable for the court proceedings will be set by the court, either in writing or in an early preparatory oral hearing that is summarised by the court in writing. The defendant will be given several weeks, or up to a few months, to file a reply writ. After the reply writ, both parties usually file one further writ. The writs should contain detailed statements of facts, offers of evidence and legal evaluations. They are crucial for the preparation of the oral hearing. The main oral hearing typically lasts only a few hours. The hearing is directed by the presiding judge, who plays an active role. Opening the hearing, the court summarises the parties' pleadings. Frequently, the court gives some indication as to its provisional assessment of the case. The court may pose specific questions to counsel for the parties, and the counsel respond to these orally. The court decides whether additional evidence needs to be gathered, for example by appointing a neutral expert or – rarely – by hearing witnesses. If a neutral expert is appointed, a further oral hearing may be held once the expert has submitted his report. The court pronounces its decision one week, or up to a few weeks, after the (last) oral hearing. It then delivers its written decision with grounds from one week, up to a few weeks, later.

Typical steps and likely timescales of infringement proceedings before the District Court of Düsseldorf can be as shown in the Table below. (The time periods indicated are based on experience and may vary depending on the actual workload of the court.)

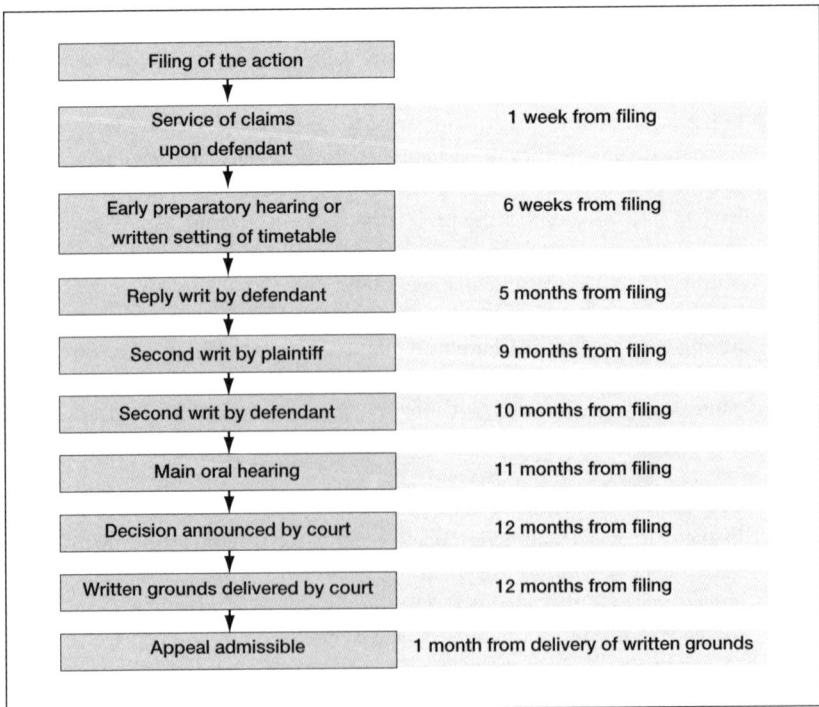

Filing of the action	
Service of claims upon defendant	1 week from filing
Early preparatory hearing or written setting of timetable	6 weeks from filing
Reply writ by defendant	5 months from filing
Second writ by plaintiff	9 months from filing
Second writ by defendant	10 months from filing
Main oral hearing	11 months from filing
Decision announced by court	12 months from filing
Written grounds delivered by court	12 months from filing
Appeal admissible	1 month from delivery of written grounds

Appeal proceedings to the competent Higher Regional Court in patent infringement litigation typically last around 15 to 18 months.

In Germany, actions on the merits for patent infringement and proceedings for patent invalidity are separate from one another. In an infringement action on the merits, there is no invalidity defence or counterclaim of invalidity against the patent in suit. The validity of the patent in suit can only be tried in separate opposition proceedings before the European Patent Office and the German Patent and Trademark Office respectively, or in separate revocation proceedings before the Federal Patent Court. Revocation proceedings are not admissible as long as opposition proceedings are admissible.

Invalidity issues can be relevant in an infringement action on the merits only in so far as the infringement court may stay the proceedings. A stay may be ordered if:

- opposition or revocation proceedings are pending against the patent in suit; and
- the infringement court is convinced that it is highly likely that the patent will be declared invalid.

However, an infringement court will basically accept the patent in suit as granted. In general, only 10 to 15% of infringement actions are stayed.

As another consequence of the split proceedings, there is no file-wrapper estoppel in German infringement proceedings.

If only after final judgment on infringement the patent in suit is declared invalid in patent revocation proceedings, the infringer can file an action for retrial of the infringement case (*Restitutionsklage*). Therefore, ultimately, a patent nullity decision will prevail over an infringement court decision finding infringement.

In the case of a patent revocation action, it typically takes about 21 months until a judgment is issued by the Federal Patent Court. Patent revocation appeal proceedings before the Federal Court of Justice last at least four years. This duration is increasing – and is one of the hot topics in German patent law (see Section 12 below).

Accelerated proceedings are not available.

5. Availability of pre-action evidence gathering/disclosure

There is no pre-action disclosure (or discovery) procedure under German law. Instead, pre-action search order proceedings have been available since a major shift in the leading case law in 2002 (the Federal Court of Justice's *Faxkarte* decision). They are now codified in the German Patent Act. German infringement courts are able to order an alleged infringer (or a third person) to present and have inspected certain objects (including corresponding documents and, if applicable, software source codes) which it has in its possession. Such search orders may be granted *ex parte*. Prerequisites for any search orders are:

- that the patent proprietor (or licensee) can show a "certain degree of probability" of a patent infringement; and
- that the patent proprietor has no other means of verifying and proving the infringement.

If a search order is granted, the respective object can be examined on the alleged infringer's premises by a court-appointed neutral expert and counsel for the patent proprietor, who can be allowed, for example, to dismantle the product or to take samples for later analysis. Both the neutral expert and counsel for the patent proprietor are sworn to secrecy vis-à-vis the patent proprietor. After the expert has delivered its findings from the inspection in a written report, the court discusses the findings in a non-public hearing with counsel for the parties. The court then decides if and to what extent the findings from the inspection may be disclosed to the patent proprietor. The decision depends on balancing the alleged infringer's interest in protection of the trade secrets which may be involved against the patent proprietor's interest in securing evidence of a patent infringement.

Search order proceedings do not oblige a party to commence substantive proceedings. To the extent that the court decides that findings from the inspection may be disclosed, German law does not bar the patent proprietor from exporting such findings to other jurisdictions.

6. Availability of interim relief (especially injunctions)

Preliminary injunction proceedings are available in patent infringement matters. In cases of particular urgency, the infringement court may issue an *ex parte* injunction within just a few hours or days after the application. As a rule, however, the court will schedule an oral hearing at short notice and render a decision within a few weeks after the application. Judgments in injunction proceedings are enforceable without security having to be furnished. The applicant can wait for up to one month to serve an *ex parte* injunction upon the infringer. An *inter partes* injunction which is granted subsequent to an oral hearing is served upon the parties by the court. The most frequently chosen district courts for preliminary injunction proceedings are Düsseldorf, Mannheim and Hamburg.

In preliminary injunction proceedings, the applicant does not have to furnish full evidence but, rather, prima facie evidence to back up the claims made in its application – for example, reports by its party-appointed experts or affidavits by witnesses. The court will examine:

- the probability of an infringement;
- the probability that the allegedly infringed patent is valid;
- the time that has lapsed since the applicant has become aware of the infringing acts; and
- the economic interests of the parties in the matter (eg undercutting prices of the accused goods).

It follows from the second point above that the separation between proceedings for patent infringement and proceedings for patent invalidity, as applicable to actions on the merits, is qualified to some extent in preliminary injunction proceedings. The infringement court will examine invalidity issues. For point three above, the applicant must show 'urgency' in terms of time. After finding out about the infringing acts, the applicant must act swiftly. Depending on the forum chosen, an application should be filed within two months.

If a party is expecting to be injuncted, it may lodge a protective brief (*Schutzschrift*) with the competent District Courts to defend itself, the main idea being to stop the injunction being issued, or at least to get an oral hearing scheduled.

On the plus side, preliminary injunctions offer the quickest way of obtaining judicial relief. If the alleged infringer contravenes an injunction served on it, it may be fined up to €250,000 (to be paid to the treasury), or sentenced to up to six months' custody for disobedience of a court order. On the con side, the applicant may face problems in furnishing prima facie evidence and is liable for any damages the alleged infringer may suffer from the injunction if the injunction is revoked on appeal.

7.   Disclosure

There is no disclosure (or discovery) procedure under German law. Instead, pre-action search order proceedings are available as described Section 5 above. Furthermore, German procedural law enables German courts to request any of the parties (or a third person) to produce specific documents which the party has in its possession and which the other party refers to. A party may object to the request for production if the respective documents contain confidential information. While the court does not have the means to force a party to produce any documents, within the scope of its free evaluation of the evidence furnished it may draw (possibly negative) conclusions from a failure of the party to do so.

8.   Evidence

8.1   How evidence is put before the court

The most relevant types of evidence in proceedings are documentary, witness and expert evidence. Courts usually rely most heavily on documentary evidence. In an action on the merits, witnesses do not give affidavits, but oral testimony. However, the court has discretion as to whether witnesses should be heard – and will usually refrain from hearing a witness.

8.2   Experts

Expert evidence is relevant only in about 10% of all patent cases. This type of evidence is given by a neutral expert who is appointed and instructed by the court. Typically, expert evidence is given in order to determine whether certain patent claim elements are featured by the allegedly infringing embodiment, where the court lacks the necessary technical background to make a decision on this aspect of proceedings. The parties may propose certain experts, but such proposals are not binding on the court. A party may also submit the opinions of experts instructed by the party, for example in order to report on certain experiments conducted and/or in order to rebut the report of a court-appointed expert. However, opinions of party-appointed experts usually carry less weight than the opinion of a court-appointed expert.

If the court decides it wishes to hear them, witnesses and experts are mainly examined orally by the court. Additional examination by counsel for the parties is practised but usually less important.

In preliminary injunction proceedings, witnesses usually give affidavits, but oral testimony is also possible. Neutral experts are not available, as the appointment of court-appointed experts would delay proceedings. A party may only submit the opinions of experts instructed by it.

8.3 Burden of proof

In patent infringement litigation, the burden of proof for establishing an infringement rests with the party whose rights are alleged to have been infringed. However, the infringing party bears the burden of proof that certain exemptions apply, for example, the experimental use exemption, or that the patent in suit is likely to be invalid. If the infringement charge is based on a patent which claims a process yielding a new product, the burden of proof shifts in part: The party who alleges infringement only has to prove that the infringing party makes the same product as yielded by the patented process. The burden is then on the infringing party to prove that it actually uses a different process.

In patent revocation proceedings, the burden of proof for invalidity of the patent rests with the plaintiff.

9. Law

The main source of German patent law is the German Patent Act. This act is harmonised with the European and Community Patent Conventions (EPC/CPC) and European Directives, such as Directive 2004/84/EC on the enforcement of intellectual property rights and Directive 98/44/EC on the legal protection of biotechnological inventions. Another important source of law is the German Act on International Patent Conventions which, among other things, regulates the interaction of German patent law and the EPC with regard to the requirement to translate European patent specifications into German if they are not published in German by the European Patent Office.

9.1 The test for patent infringement

Under the German Patent Act it constitutes a patent infringement commercially to manufacture, offer for sale, place on the market, use or possess/import into Germany for the aforementioned purposes a patented product without the patent proprietor's consent. Where the subject matter of a patent is a process, it constitutes a patent infringement to use this process without the patent proprietor's consent, or to offer this process for use if it is evident that the use is not allowed without the patent proprietor's consent; or to use, offer for sale, place on the market, or possess/import for the aforementioned purposes a product obtained directly by this process.

Under the German Patent Act, it constitutes a contributory patent infringement to offer for sale, or deliver without the patent proprietor's consent in Germany, means which relate to an essential element of the patented invention if it is evident that the receiving party of the offer or delivery will utilise the means to use the invention in Germany.

To assess whether a patent is infringed, the court will first construe the patent claims without referring to the accused goods or process. When interpreting the language of a patent claim, intrinsic evidence (ie evidence found within the patent document itself) is more persuasive than extrinsic evidence, for it is acknowledged that a patent applicant can be his own lexicographer and assign new meanings to generally known terms. Subclaims, descriptions and drawings of the patent are be taken into consideration when construing patent claims. However, as rule, the content of subclaims, description or drawings does not justify a limitation of a claim, if this limitation is not reflected in the language of the claim itself.

Generally there is no file history estoppel. The prosecution history of a patent, including the grounds of a decision in patent opposition or revocation proceedings, may not be taken into consideration when construing patent claims. An exception based on the general precept of good faith applies only:

- if the patent proprietor has defended the validity of its patent in patent opposition or revocation proceedings by declaring a limitation on the scope of protection of its patent; and
- if the alleged infringer was party to these opposition or revocation proceedings.

The court may appoint a neutral expert to report on the understanding of the language of the patent claims by a person skilled in the art on the priority date of the patent. Yet, ultimately, the court must still assume responsibility for interpreting the patent claims.

A patent can be infringed either literally or under the doctrine of equivalents. An infringement is literal if each and every claim element – as it is construed – is featured by the accused goods or process.

Leading case law from 2002 (the Federal Court of Justice's *Schneidmesser I* decision and others) has established that an infringement under the doctrine of equivalents is to be assessed in the following three-step test. First, one examines whether the accused goods or process solve the technical problem addressed by the claimed invention by means which are modified vis-à-vis those of the claimed invention (otherwise there would be literal infringement), yet have the same technical effect. If this is the case, it is then determined whether a person skilled in the art, based on his general knowledge and skills (ie without inventive efforts), would have been able to identify the means of the accused goods or process as having the same technical effect. If so, it is finally asked whether these considerations by the person skilled in the art are geared toward the meaning of the patent claim in such a way that the person skilled in the art would consider the accused goods or process as a technical solution equal to a literally infringing good or process. If all three questions are answered in the affirmative, there is infringement under the doctrine of equivalents.

According to leading case law from 1986 (the Federal Court of Justice's *Formstein* decision), an exception applies to the lack of an invalidity defence in a patent infringement action. If an infringement is not literal, but occurs only under the doctrine of equivalents, the accused goods or process are exempted if their features were obvious from the prior art relevant for the patent in suit.

9.2   Assessment of patentability

According to German law, patentability requires:

- an invention;
- the invention must be novel;
- it involves an inventive step; and
- it is susceptible of industrial application.

The term invention is defined as a technical teaching, that is, an instruction to solve a technical problem by employing technical means and/or forces of nature, not by employing only the human mind. The term does not include, *inter alia*, the following:

- discoveries which only enrich human knowledge, but not the human technical skills;
- aesthetic creations;
- business methods; and
- computer software.

However, the aforementioned subject matter is excluded from patentability only as such. That is, patents may not be granted for computer software as such, but may be granted for computer-implemented inventions (eg for computer software controlling or interacting with technical devices, or for the computer-implemented processing of certain steps of a procedure which solves a technical problem). The exact boundaries between non-patentable computer software and patentable computer-implemented inventions are a hot topic in German patent law.

Patents may be granted for inventions that relate to biological material or processes that yield or use biological material. Biological material which is isolated from its natural environment by means of a technical process can be a patentable invention, even if such material was previously present in nature. The patentability of human stem cells has not been resolved under German law. However, the question of the patentability of human stem cells, in particular with regard to concerns based on the *ordre public*, is currently pending before the Enlarged Board of Appeal of the European Patent Office under file number G 2/06. The expected decision is likely to have a strong impact on the evaluation of the patentability of human stem cells under German law. Patents for human genetic sequences can only be granted if the patent application discloses a specific technical or biological function of each claimed sequence. If a patent is granted for a genetic sequence that has the same composition as a naturally occurring human genetic sequence, a specific technical or biological function of the sequence must be included in the respective patent claims. One hot topic under German patent law is whether this constitutes an exception to the principle of absolute protection of substance patents and, if so, whether such exception conforms to the international Agreement on Trade-Related Aspects of Intellectual Property Rights (TRIPS).

Patents may not be granted for processes for the cloning of human beings, for the use of human embryos for commercial purposes, for processes for modifying the genetic identity of the germ line of a human being, or for processes for modifying the genetic identity of an animal, if such process is likely to cause the animal to suffer without medicinal benefits for man.

In line with the generally accepted *novelty test*, an invention is novel if it differs from the disclosure of every single piece of prior art in at least one claim feature. An invention is still novel if all individual claim features were known from the prior art when taken together, but not the inventive combination of these features. The state of the art is constituted by public written or oral description, or other public disclosures like public use. The novelty test is absolute; that is, an invention must be novel in respect of prior art anywhere in the world, independent of the inventor's knowledge of the art.

For the novelty test (but not for the test for inventive step), the content of German, European and international patent applications with Germany as the designated state are considered as comprised in the state of the art not only if they have been published before the priority date of the patent to be examined, but also if they have been filed before the priority date and published on or after the priority date.

Under German law, there is generally no grace period for novelty. The inventor's own prior public disclosure of the invention constitutes novelty-destroying prior art. An exception to the absolute novelty requirement applies only to certain prior disclosures of an invention that are based on:

- an evident abuse (breach of duty) in relation to the applicant; or
- a display of the invention by the applicant at exhibitions that are officially recognised for novelty protection (which in essence are certain international trade exhibitions and world fairs).

An invention involves an inventive step if – considering the prior art – it was not obvious to a person skilled in the art. The non-obviousness is assessed from an *ex ante* perspective, based on the knowledge and skill of a hypothetical person skilled in the art on the priority date of the patent. Such hypothetical person can be a team of several experts. For example, in regard to pharmaceutical patents, such team may include analytical chemists, medical scientists and pharmacists.

The test for inventive step involves the following questions: First, based on the relevant prior art and its possible combinations, which steps did the person skilled in the art have to take in order to arrive at the inventive solution? Secondly, was the person skilled in the art induced to direct its considerations at the inventive teaching? Thirdly, what are the arguments for and against the contention that, by means of such considerations, the person skilled in the art would have arrived at the inventive solution? In addition, the following indications of inventive step may be taken into consideration: overcoming of technical difficulties; technological gap vis-à-vis prior technical solutions; overcoming of false preconceptions among persons skilled in the art; solution of a technical problem which has existed for considerable time; and commercial success in marketing embodiments of the invention.

An invention is susceptible of industrial application if it can be made or used in any kind of industry, including agriculture. Almost every invention of a product is susceptible of industrial application. Methods for treatment of the human or animal body by means of surgery or therapy and diagnostic methods practised on the human or animal body are not susceptible of industrial application. This exception does not apply to products, in particular substances or compositions, for use in such methods.

Under German law, a patent application must disclose the invention in a manner sufficiently clear and complete for it to be carried out by a person skilled in the art. This requirement has to be fulfilled at the time the patent application is filed.

Sufficiency of disclosure does not have to be established by the patent claims alone. Also relevant is the understanding by a person skilled in the art of the description, drawings, technical interrelations of features and the technical problem to be solved by the invention, as well as the general technical knowledge present in the art. A disclosure is insufficient if the person skilled in the art can practise the teaching of the invention only with great difficulty and after overcoming prior failures.

10. Available remedies

The most common remedies are injunctions (cease and desist) and damages. A claim for damages requires that an infringer acts with deliberate intent or negligence. Negligence is generally assumed because companies – at least producers and importers of goods – have an obligation to be informed about the patent situation in the relevant sector.

A patent proprietor (or licensee) generally does not know the extent to which the patent in suit has been wrongfully used at the time when it lodges the complaint, so that it is unable precisely to calculate the amount of damages it may claim. For this reason, the patent proprietor can include claims for information and rendering of accounts against the infringer. Among other things, the infringer has to provide a detailed list of the turnover generated from sales of the relevant goods and – broken down according to the individual cost factors – its actual costs and profits. It also has to provide the names and addresses of customers.

Usually, the plaintiff will request a declaratory judgment on damages to the effect that the defendant must compensate the plaintiff for the losses it has sustained, and will sustain in future, as a result of the infringing acts. The amount of damages has to be determined in separate proceedings, after the infringer has provided the relevant information and accounts.

10.1 Calculation of financial remedies

Damages can be calculated in three ways:

- lost profit;
- lost reasonable royalties; and
- surrender of the profit generated by the infringer.

Since a major shift in the leading case law in 2000 (the Federal Court of Justice's *Gemeinkostenanteil* decision), overheads may now be deducted from the infringer's turnover only if, and to the extent that, they are directly and exclusively allocable to the infringing goods. This has strengthened the patent owner's position.

Even if claims for damages are statute-barred (a patentee may only claim damages suffered during the three calendar years preceding awareness by the patentee of the infringing acts), or if there is no fault, the court may grant a claim for unfair enrichment, which is usually calculated according to lost reasonable royalties.

Further remedies are claims for destruction of accused goods and – under the influence of the European Enforcement Directive 2004/48/EC – claims for recall from the channels of commerce of the accused goods and for publication of a judgment at the infringer's expense. Claims for information by the party who is claiming infringement can even cover the infringing party's banking documents.

In theory, criminal charges are also available against an infringer who acts with deliberate intent. Yet criminal proceedings have not gained major practical importance.

10.2 The effect of an appeal on enforcement

An infringement court judgment is provisionally enforceable against the provision of security – even if the judgment is appealed by the alleged infringer. An appeal has no suspensive effect. The security is usually provided in the form of a bond by a German bank. The amount required is fixed by the court and usually equals or slightly exceeds the litigation value of the claims to be enforced.

11. Costs

The costs of patent litigation are case specific. An 'average' patent infringement action on the merits may cost around €50,000 to €80,000 for the first instance, including court fees and the fees of a party's lawyer and patent attorney.

The losing party bears the court fees and, at least to a limited extent, has to reimburse the legal costs of the winning party. The costs to be reimbursed by the losing party are governed by statute. They are calculated in conjunction with the litigation value (*Streitwert*) which is determined by the court, based on the plaintiff's economic interest in the claims pursued in the action. For example, if the court determines the litigation value to be €1 million, the court fees will be about €13,400, and reimbursable costs of the winning party's lawyer and patent attorney will be about €11,300 each.

12. Hot topics

It is to the benefit of patent proprietors that actions on the merits for patent infringement are separated from proceedings for patent invalidity benefits, for invalidity issues rarely delay the enforcement of patents. However, in preliminary injunction proceedings, the separation is qualified to some extent, as the court will examine the probability that the allegedly infringed patent is valid. Making recent legal news has been the fact that in a proceeding in 2008 which related to drugs containing the active ingredient Olanzapine, the Higher Regional Court of Düsseldorf further strengthened the position of patent proprietors in preliminary injunction proceedings. The court assumed that the grant of a preliminary injunction was not per se excluded (due to lacking probability of patent validity) if the patent in suit had been revoked in first instance.

As indicated in Section 4 above, the duration of patent revocation appeal proceedings before the Federal Court of Justice has become problematic. There have been proposals to speed up these proceedings by streamlining them (ie approximating them to a purely legal review).

In principle, activities which take place outside Germany do not support a charge of infringement of a German patent, or of the German part of a European patent. However, under certain conditions, German courts may assume liability for aiding a patent infringement in Germany, even based on acts performed completely outside Germany. For example, in its *Herzkranzgefäß-Dilatations-Katheter* decision of 2003, the District Court of Düsseldorf accepted liability of a foreign producer who delivered its products only to a foreign distributor, due to the fact that the producer knew that the distributor distributed the (infringing) products in Germany.

In general, the conditions and extent of liability for contributory patent infringement have become a hot topic. In particular, since 2001, a series of Federal Court of Justice decisions (and legal papers) have addressed details of such liability.

In the last few years, the availability and extent of a FRAND defence to patent infringement have also become topical. The FRAND defence relates to an obligation which is often required by industry groups that set common standards for particular technologies (like the MPEG LA patent pool) in order to ensure compatibility and interoperability of devices manufactured by different entities. Members of the industry group are obliged to license patents under fair, reasonable and non-discriminatory terms (hence the mnemonic FRAND). In its 2006 *Videosignal-Codierung I* decision, the District Court of Düsseldorf held that the user of a patented invention who has – based on a FRAND obligation of the patent proprietor and antitrust law – a claim for a licence under a patent also has a defence to patent infringement.

Under German law of employees' inventions, the employer may, within a preclusion period, declare that certain inventions made by its employees shall be transferred to the employer. Without such declarations, employees' inventions essentially remain the sole property of the employee inventors. In its controversial *Haftetikett* decision of 2006, the Federal Court of Justice decided that under certain conditions the preclusion period for the transfer of an employee's invention can begin to run on the date of the patent application, even if the employee inventor did not correctly notify the employer of the invention. This case law considerably impedes the transfer of employees' inventions to the employer. In reaction to the *Haftetikett* decision, there has been a proposal to adapt the law of employees' inventions. The transfer of employee's inventions shall be facilitated by feigning of a respective declaration of transfer by the employer.

Germany summary

- Germany has specialised courts for patent litigation. The courts have an excellent reputation for their expertise and the quality of their decisions. The German courts hear more patent infringement cases than the courts of any other European country.
- In some respects, German patent law tends to favour the patent proprietor. One such aspect is the fact that the separation between an action on the

merits for patent infringement and proceedings for patent invalidity benefits patent proprietors. Invalidity issues rarely delay the enforcement of patents. The split proceedings limit the opportunity for the defendant to run 'squeeze' arguments (ie to limit the scope of the patent claims in an infringement action).

- German courts work swiftly. In the case of a suit for patent infringement, it typically takes 8 to 12 months until a judgment is issued by the court of first instance. Preliminary injunctions are obtainable within a few hours or days in *ex parte* proceedings, or within a few weeks if an oral hearing is scheduled.
- An average patent infringement action on the merits may cost around €50,000 to €80,000, including court fees and fees for one party's lawyer and patent attorney. The losing party bears the court fees and, at least to a limited extent, has to reimburse the legal costs of the winning party.
- In German litigation, there is no disclosure. Instead, pre-action search order proceedings are available, and parties can be requested to produce specific documents which the other party refers to.

Greece

George A Ballas
Loukia K Papas
Ballas, Pelecanos & Associates

1.   Lawyers

Patent litigation in Greece is generally handled by lawyers/attorneys who specialise in the field of industrial and intellectual property. More specifically, they are experts with regard to patent litigation.

There is no distinction in Greece between barristers and solicitors.

Industrial property issues belong to the exclusive competence of a special Court Department, which has existed and been functioning since May 2006 in the Courts of Athens and Thessaloniki. Therefore, patent litigation can only occur before these two jurisdictions. Lawyers/attorneys handling patent litigation in a jurisdiction different from the one to which they are appointed require a local lawyer to provide them with the necessary authority in order that they may appear before the competent court.

2.   The court system

The civil court system in Greece comprises the Court of First Instance, the Court of Appeals and the Supreme Court of Cassation (*Areios Pagos*). A regular lawsuit is filed with the Court of First Instance, the final decision of which can be appealed before the Court of Appeals within an exclusive deadline of 30 days from the date on which the service of the final decision to the adverse party has taken place. The decision issued by the Court of Appeals can then be disputed before the Court of Cassation within 30 days starting from the service of the Court of Appeals decision to the adverse party. This appeal, however, can take place only for reasons relating to the proper interpretation and application of the relevant legal provisions by the ruling court. In contrast, both the Court of First Instance and the Court of Appeals deal with the dispensation of substantive justice.

As stated above, industrial property issues belong to the exclusive competence of a special court department functioning in the Courts of Athens and Thessaloniki. This department consists of three judges, specialising in, or having requisite experience with regard to, patent litigation as well as commercial law in general. These judges lack technical backgrounds, and for this reason they will usually make use of expert reports on technical issues arising during patent litigation. It is at the discretion of the ruling court to take note of decisions issued in other jurisdictions. However, it is evidently more likely that the court will take into account a decision that has been issued on an ad-hoc basis, or at least on similar patent-related issues.

Under Law 1733/1987 "Technology transfer, inventions, and technological innovation", currently in force, the owner of a patent can proceed with various actions before civil justice. Moreover, depending on the facts of each particular case, possible grounds for civil actions lie under the Laws on Unfair Competition, along with the Civil Code provisions. More specifically:

- *Nullification action*: According to article 15 of Law 1733/1987, the patent will be declared null by the court if:
  - the owner of the patent is not the inventor or his assignee or beneficiary according to article 6, paras 4, 5 and 6 of Law 1733/1987;
  - the invention is not patentable in accordance with article 5 of Law 1733/1987;
  - the description attached to the patent is insufficient for the invention to be carried out by a person skilled in the art; or
  - the subject matter of the granted patent extends beyond the content of the protection, as requested in the application.

  If the nullification is brought before the court only against part of the invention, the patent is restricted accordingly.

- *Action for cessation of the infringement and prohibition against future infringement*: According to article 17 of Law 1733/1987, in the event of existing or threatened infringement of the patent, its owner has the right to demand the cessation of the infringement and a prohibition against any future infringement.

- *Action for damages*: According to article 17 of Law 1733/1987, in the event of intentional infringement of a patent, its owner is entitled to demand restitution of the damage, or return of the benefits derived from the unfair exploitation of the invention, or the payment of an amount equal to the value of a licence for such exploitation. The same rights are granted to the beneficiary of an exclusive licence, to any party with a right over the invention, and to any party who has filed a patent application. In the latter case, the court may postpone the trial procedure of the case until the relevant patent has been granted. It should be noted that if the invention relates to a process for the manufacture of a product, each product of the same nature is presumed to have been manufactured according to the protected process. The above-mentioned rights will be prescribed after five years have elapsed from the date on which the owner of the patent became aware either of the act of infringement, or of the person liable to give compensation and of the damage caused by the infringement, and certainly after 20 years have elapsed since the infringement took place. If the defendant is found to have infringed the relevant rights, the court may order the destruction of the products manufactured in violation of Law 1733/1987. Alternatively, the court may order that the products or a part thereof be rendered to the plaintiff for his total or partial compensation, at the request of the plaintiff.

- *Claim of ownership*: According to article 6 para 9 of Law 1733/1987, the beneficiary of the invention may, if a third party has filed without consent a patent application relating to his invention or to essential constituents thereof, apply for legal recognition of his rights resulting from the patent application or, if the patent has already been granted, his rights resulting from the patent. This action must be brought before the court within a period of two years from the date of publication of the summary of the patent in the Industrial Property Bulletin. This term does not apply if the patentee is aware of the right of the claimant at the time of grant or assignment of the patent. A summary of the irrevocable decision stating the acceptance of this action will be recorded in the Patents Register. The licences and all other rights which have been granted on the patent will be considered null as from the date of this record. If the defeated litigant and third parties had exploited the invention in good faith, or had proceeded with the necessary preparations for the relevant exploitation, they may request from the recognised beneficiary a grant against compensation of a non-exclusive licence for a reasonable period of time.

- *Recognition action*: Based on article 70 of the Civil Code, the claimant can file an action before the competent court claiming recognition of the existence or non-existence of an exclusive right to exploit a specific patent or of the obligation of the defendant to compensate the owner of a patent, without specifying the exact amount of such compensation.
- *Injunction*: Under article 731 of the Civil Procedure Code, in the event of actual or threatened infringement of a patent, its owner has the right to file before the competent court an injunction application, demanding cessation of the infringement and a prohibition against any future infringement.

All the above-mentioned legal actions, with the exception of the injunction application, can only be discussed if the parties have attempted to reach an extrajudicial settlement according to article 214A of Civil Procedure Code.

Warning letters, before resorting to the filing of an action, are considered to be a good idea.

3.   Procedure and timescale of proceedings

After filing the regular lawsuit, the parties must submit their pleadings and evidentiary material in support of their arguments 20 days before the hearing date. The hearing date is usually fixed eight to nine months from the day on which the lawsuit is filed.

The mutual reply to the litigant parties' arguments takes place with the filing of an addendum, 15 days before the hearing date. It should be emphasised that where pleadings or an addendum are not filed in due time, they will not be taken into consideration by the court. Finally, within eight working days after the hearing date, the parties file the evaluation of any testimonial statements. The issuance of the decision by the Court of First Instance usually takes place within four to six months after the hearing date.

A party with a legitimate interest can file an appeal within 30 days following service of the decision if the party is resident in Greece, and 60 days if the party is resident abroad. The pleadings, along with the evidentiary material of the parties, must be submitted to the court before commencement of the procedure, and the filing of the addendum must take place within three days of the hearing date, which is usually fixed within four months of the filing of the appeal. The issuance of the decision usually takes place four to six months after the hearing date.

The hearing date for injunction applications is usually fixed within two months after its filing. The pleadings and evidence of the parties must be filed within two or three days of the hearing date, following a decision of the residing judge at the hearing of the case.

The parties can submit an application asking for the fixing of an earlier hearing date, and citing important reasons for such a request. The judge examines the application and decides whether to accept or reject it. In practice, an earlier hearing date is set when the disputed claim is based wholly or partly on unfair competition law. The hearing date is fixed six months after the filing of the regular lawsuit, so as to prevent the prescription of the claim based on unfair competition.

If opposition proceedings have been initiated before the European Patent Office (EPO) under article 99 of the European Patent Convention, the competent Greek court will usually postpone the trial of the case until the EPO decision on the opposition filed is issued and notified to the litigant parties. This postponement is appropriate, since the opposition applies to the European patent in all the contracting states in which the patent has effect and may even lead to the revocation of the European patent, in which case the European patent will be deemed not to have had, from the outset, the effects specified in articles 64 and 67 of the European Patent Convention. More specifically, within nine months of the publication of the mention of the grant of the European patent in the *European Patent Bulletin*, any person may give notice to the European Patent Office of opposition to that patent. The opponents are parties to the opposition proceedings, as well as the proprietor of the patent. If the Opposition Division of the EPO is of the opinion that at least one ground for opposition prejudices the maintenance of the European patent, it will revoke the patent. Otherwise, it will reject the opposition and the European patent will remain valid in all the contracting states in which it has effect.

4.   Availability of pre-action evidence gathering/disclosure

As already stated, in the event of intentional infringement of a patent, its owner who suffered damage is entitled to demand restitution of the damage or return of the benefits derived from the unfair exploitation of the invention or the payment of an amount equal to the value of the licence for the unfair exploitation. The patent owner is often unable to estimate the extent of the damage suffered due to the unauthorised use of his/her patent and therefore will not be in a position to calculate accurately the amount to claim in terms of an action for damages. For this reason, the rights holder may file a lawsuit against the infringer claiming the disclosure of all relevant information. The requested information may concern the quantity of the products that were illegally circulated by the infringer, as well as the profits that the latter made due to the illegal actions. Both the nature and the extent of the information to be granted are specified in each particular case according to the needs of the rights holder, as well as the reasonable interests of the infringer, who cannot in any event be forced to reveal the commercial secrets of his business. The usual practice entails the appointment by the court of a financial expert, who will inspect the infringer's official books.

The rights holder may file a regular lawsuit before the competent Court of First

Instance. However, it is also possible to file an injunction application to the same end. In the latter case, the preconditions mentioned in Section 5 below must be fulfilled in order for the application to be accepted. The plaintiff must prove his legitimate interest in order to obtain sensitive information, as the courts are not usually willing to authorise a third party's access to the official books or business practices of an alleged patent infringer, especially in the context of a speedy injunction procedure that may not be followed by the filing of a regular lawsuit.

The injunction decision retains its validity until the issuance of a final decision by the court on the regular lawsuit in the same matter, provided the lawsuit is filed and served on the defendant within 30 days from the issuance of the injunction decision.

It should be noted that the information granted by virtue of a court decision on such a lawsuit/injunction application can only be used in the context of a dispute between the same litigant parties, either in Greece or in any other competent jurisdiction. The information provided to the rights holder cannot be used in any other current or future dispute between himself and a third party. Additionally, the rights holder may not use the information for the purpose of unfair competition against the alleged infringer, or for any other generally unlawful use.

5. Availability of interim relief (especially injunctions)

Preliminary measures will often involve the filing of an injunction application against an infringer. If the applicant is successful at the relevant hearing before the competent First Instance Court, an injunction will be granted. The main conditions for the acceptance of an injunction application are the following:

- an urgent need for the grant of award provisional measures; or
- imminent danger to the interests of the plaintiff patent owner/beneficiary as a result of the alleged infringement.

The existence of these factors must be shown by the plaintiff. Timing is therefore crucial in this procedure. Thus, an injunction procedure must be filed within a short time after the patent owner/beneficiary becomes aware of the infringement.

On filing an injunction application, the plaintiff rights holder may request a judge, appointed by the court for this purpose, to grant a temporary restraint order (TRO) with the intention of securing the preservation of the rights holder's interests prior to the grant of provisional measures. The validity of the TRO is usually extended by the court at the hearing of the injunction application up to the issuance of the decision. It is at the discretion of the judge to issue such an order, except for cases involving infringement of copyright where its issuance is compulsory. A TRO may be issued *ex parte* (ie without notice to the defendant), but the practice is that the judge will summon the defendant at its hearing.

Significant advantages of the injunction procedure are:

- It is not necessary to provide full proof of the rights holder's claim. It is enough that the rights holder is able to offer persuasive arguments as to the claim and to show that there is a substantial likelihood that his rights are being infringed;
- The injunction decision to be issued is non-appealable and, in spite of its

temporary nature, enforceable;

- The court has the authority to order all and any kind of provisional measures which are appropriate for the protection of the rights of the plaintiff, according to its discretion, and even if those measures are not provided in an explicit way by the law. Such provisional measures are:
  - the imposition upon the infringer of the obligation to cease and desist from the infringement (illegal use and/or production and/or distribution of the infringing products) and withdraw temporarily any infringing products from the market;
  - the detailed inventory of infringing items in the possession of the infringer or their temporary confiscation and sequestration;
  - the audit of the commercial records and books for the gathering of data as to the infringing items (conducted by a technical expert appointed by the court);
  - the threat of penalties (both monetary fine and personal incarceration) in the event that the court's orders are ignored and for each violation of its provisions; and
  - publication of the order of the injunction decision in the press.

The main disadvantage of the injunction procedure is its extremely tight timeframe, within which the plaintiff may not be able to persuade the ruling judge as to the urgent need for the grant of a preliminary protection and the imminent danger threatening his/her interests.

6.   Disclosure

By virtue of articles 450 to 451 of Greece's Civil Procedure Code, in the context of a pending patent litigation every litigant or third party is obliged to disclose and demonstrate the documents in his possession that may be used as evidence in the court, unless there is a justified reason for not doing so. The preconditions for the application of such disclosure obligation are:

- the application of a litigant party submitted before the ruling court at every stage of the trial;
- the exact specification of the documents to be revealed in addition to their content;
- the possession of the documents by the opposing party or third party;
- the admissibility of the documents to the court as probative evidence; and
- the legitimate interest of the claiming party.

The disclosure is limited to the extent necessary for the clarification of the issues carefully and adequately cited in the relative application before the court and cannot be expanded to information beyond the scope of the pending dispute. The documents to be disclosed can only be used in the context of the pending dispute, as well as in any other dispute between the same litigant parties, taking into account the provisions of the law pertaining to the protection of confidential information/business secrets of the parties.

7. Evidence

The probative evidence that can be taken into consideration by the ruling court include:

- confessions;
- autopsies;
- experts' reports;
- documents;
- witnesses;
- examination of the litigant parties;
- litigant parties' oaths (ie the affirmation by a litigant, of the truth or falsity of specific facts under oath); and
- judicial presumptions (article 339 of the Civil Procedure Code).

It is at the court's discretion to evaluate and assess the probative value of the produced evidence, which may be in the form of documentary evidence (eg expert reports, written arguments or witness statements), or live witnesses giving oral testimony before the ruling court. Live witnesses (including experts) are subject to cross-examination.

As far as experts are concerned, according to article 368 of the Civil Procedure Code, the ruling court may appoint by its decision one or more of the experts included in the relative official experts' list available in every court, if it deems it necessary for certain technical or scientific issues to be clarified/estimated. In the event that a litigant party explicitly requests that an expert is appointed by the court, the court is legally obliged to proceed with the appointment of an expert, provided that it also reckons that special technical or scientific knowledge is required for certain issues to be resolved. The report drafted by the officially appointed experts is then submitted to the ruling court. The court is free to assess and evaluate the outcome of the technical or scientific research conducted when arriving at its decision. In most cases, however, the final decision issued in patent litigation will be based on the outcome of such a report.

Apart from the official procedure followed by the ruling court, as described above, the parties may, on their own initiative and at their own expense, resort to the appointment of one or more experts of their free choice and acquire a technical experts' report or an ad-hoc legal opinion that is then submitted to the ruling court as supporting evidence. These independently drafted reports or legal opinions can be judged by the court as it sees fit. Since it is common practice for the litigant parties to produce and submit such experts' reports to the court, the court will usually base its decision on the outcome of its own official expert report.

8. Law

According to article 10 of Law 1733/1987, a patent confers upon its owner, whether a natural person or legal entity, the exclusive and time-limited right productively to exploit the invention in any way legally feasible and financially desirable, as well as forbidding any third party from exploiting the invention or importing the products protected by the patent without the prior consent of the owner.

In the event of patent infringement, the extent of the protection conferred by a national or European patent or patent application is decisively determined by the accompanying description and drawings as well as the relative search report. The description and the drawings are used to interpret claims, but the latter along with the way in which they are outlined/phrased remain the main criteria for the acceptance or rejection of a specific patent infringement lawsuit. As described above, the ruling court may have recourse to one or more experts, who will review the relevant patent documentation, outline the applicable technical rule described in the claims which describe the essence of the invention, and then proceed with the comparison between such technical rule and the allegedly infringing national or European patent application/method or final product, so as to decide whether or not a patent infringement has actually taken place.

The doctrine of equivalence is often invoked in patent cases; however, the courts tend to accept it only as a means of interpretation or specification of the patent claims and not in order to expand a patent's ambit as set out by the relevant claims. According to the definition given to the doctrine of equivalence by the WIPO (World Intellectual Property Organisation) experts' committee, the claims of a patent cover not only the technical means specifically mentioned/described therein, but also any equivalent means. The applicability of the doctrine of equivalence deters the abbreviation of inventors' rights as well as the usage of the inventive idea by third parties through insignificant but equivalent technical, verbal or structural variations of the inventive idea. So far as this variation consists of an equivalent solution to the invention and is at the same time covered by the core of the inventive idea, the infringement is affirmed by the court, even if the grammatical wording of the patent does not cover such variation. On the other hand, the applicability of the doctrine of equivalence for the purpose of expanding the protection granted through the grammatical wording of the patent claims can only be the exception to the rule. The courts make use of this doctrine in order to specify the content of article 10 of Law 1733/1987 (which provides for the patent owner's right to forbid any third party from productively exploiting the invention) and to grant fair protection to the patent owner. This doctrine cannot, however, lead to the unfair expansion of the exclusive rights granted to the patent owner, in a way that might unreasonably affect the rights of third parties to operate in the same market.

As regards assessment of patentability, particularly in the context of a nullification action, according to article 5 of Law 1733/1987 patents will be granted for any inventions which are new, involve an inventive step and are susceptible of industrial application. The invention may relate to a product, a process or an industrial application.

An invention will be considered new if it does not form part of the state of the art. The state of the art is held as anything made available to the public anywhere in the world by means of a written or oral description or in any other way, before the filing date of the patent application or the date of priority. Patents will also be granted for an invention which has been disclosed no earlier than six months before the filing of the patent application, if the disclosure was due to:

- an evident abuse of the rights of the applicant or his legal predecessor; or

- the fact that the invention was displayed at an officially recognised international exhibition falling within the terms of the convention on international exhibitions signed in Paris on November 22 1928 and ratified in Greece by Law 5562/32. In this case, when filing the application, the applicant should state that the invention has been so displayed and should file the relevant supporting certificate. This disclosure does not affect the novelty of the invention.

In order for the ruling court to assess the novelty of a patent, it will resort to the appointment of experts, who will then undertake the necessary research in the technical databases available in every patent office around the world, most importantly the EPO and US Patent Office databases, and will submit the drafted report to the court for its final decision. According to Greek case law, an invention pertaining to the production of a product is considered to be novel if the product is essentially different (imports new characteristics) in comparison to similar products, whereas an invention pertaining to the production of a technical result is considered to be novel if it entails significant improvement of an already known result or due to its remarkable originality, irrespective of the fact that this improvement may affect only the production means or only the result, or the reduction of production costs, or all of the above.

An invention is considered as involving an inventive step if, having regard to the state of the art, it is not obvious to a person skilled in the art. The inventive step may result from indications, such as the provision of an answer to a long-existing need, the commercial success of the patent, or its unexpected outcome. An invention involves an inventive step, when a person skilled in the art considers it to represent progress over the existing state of the art. This effectively means that the invention can only be considered as involving an inventive step if an average person skilled in the art could not proceed with the solution to a problem using the (former) state of the art.

An invention will be considered as susceptible of industrial application if its subject matter may be produced or used in any sector of industrial activity (not only commercial use).

The ruling court will usually resort to the appointment of technical experts, in order to assess any of the above elements necessary for the validity of a patent and the protection of the inventive idea. However, the patentability of certain types of subject matter, such as biotech or software inventions, is treated by both applicable legislation and the ruling courts in a particular way.

More specifically, as far as the patentability of biotechnological inventions is concerned, Greece has incorporated Directive 98/44/EC "on the legal protection of biotechnological inventions" through the issuance of Presidential Decree 321/2001. According to this decree, inventions which are new, involve an inventive step and are susceptible of industrial application will be patentable even if they concern a product consisting of or containing biological material, or a process by means of which biological material is produced, processed or used. Biological material which is isolated from its natural environment or produced by means of a technical process may be the subject of an invention even if it previously occurred in nature. However, the human body, at the various stages of its formation and development, and the simple discovery of one of its elements, including the sequence or partial sequence of a gene, cannot constitute patentable inventions. Moreover, inventions will be considered unpatentable where their commercial exploitation would be contrary to *ordre public* or morality. In particular, the following will be considered unpatentable:

- processes for cloning human beings;
- processes for modifying the germ line genetic identity of human beings;
- uses of human embryos for industrial or commercial purposes; and
- processes for modifying the genetic identity of animals that are likely to cause them suffering without any substantial medical benefit to man or animal, and also animals resulting from such processes.

With regard to software inventions, according to article 5 para 2c of Law 1733/1987, computer programs are not regarded as inventions and cannot be protected via a patent. In Greece, software inventions are protected by virtue of unfair competition law as well as Law No 2121/1993 on intellectual property. The preparatory material used for the completion of the program can also be protected; however, Law 2121/1993 does not cover the ideas and principles on which any element of the program is based. It must be noted that many jurists in Greece favour the patentability of software inventions if these inventions combine a theoretical rule with a technical result (eg an electronic device or a part of a machine). According to this view, a software invention is of a technical nature when it entails the configuration of the data-processing device or the immediate use of basic parts thereof.

9. Available remedies

The remedy provided in cases of patent rights infringements is the filing of a regular lawsuit against the infringer. This is a standalone remedy, which does not have to be supported by any preliminary action; however, it will ideally be combined with an injunction application.

In the event of a regular action for damages, the plaintiff will usually make a claim for permanent cessation of the infringement and a prohibition against any future infringement. If the court finds against the defendant, it may, in addition to compensation for damages suffered, order the destruction of the products manufactured in violation of the dispositions of Law 1733/1987. Instead of destruction, the court may order that the products or a part thereof be rendered to the plaintiff for his total or partial compensation, upon request of the latter. Additionally, at the request of the winning litigant party, the court may order the publication of its decision in the daily press.

As regards financial remedies, if the infringement was intentional, the patent owner is entitled to demand restitution or return of the benefits derived from the unfair exploitation of the invention, or the payment of an amount equal to the value of a licence for such exploitation (article 17 para 2 of Law 1733/1987). This article aims to aid the patent owner in relation to the estimation of the damages suffered by granting him the discretionary power to choose any of the three available and alternative calculation methods. The plaintiff must set out in the pleadings to be submitted before the competent court an extremely accurate, detailed and specific calculation of the damages incurred. The patent litigation will then proceed on the basis of the partial or total acceptance/dismissal/adjustment of the requested sums.

A claim for the award of moral damages may also be included in the regular lawsuit. If appropriate evidence is offered in support, and depending on the facts of the case, the award of a reasonable sum to the patent owner may be granted on this ground.

It should be noted that Enforcement Directive 2004/48/EC providing for additional remedies has recently been integrated into the Greek national legislation to become part of Law 2121/1993 in relation to intellectual property, thus not being applicable in patent litigation.

The court procedure involves witness testimonies before the court and the parties are obliged to submit pleadings and supporting documentation (as set out in Section 3). The court will then issue a final decision, which is subject to appeal. The deadline within which the parties may appeal the decision of the competent Court of First Instance suspends the enforceability of such decision.

If an appeal is filed, any Appeals Court decision issued is final and enforceable, unless a further appeal is filed with the Supreme Court (though jurisdiction is here limited only to legal review) and an order of suspension is issued.

10. Costs

It is difficult to estimate the cost of a patent action, since it varies depending on the nature of the patent involved (eg pharmaceutical patent cases are always more complicated), the number of experts appointed in order to provide the ruling court with the necessary explanatory technical or scientific reports, the way in which the infringement took place and the complexity of both the legal and financial issues raised. For example, in an important patent case concerning pharmaceutical products the total cost for the injunction procedure can reach €40,000 to €50,000, whilst an increase of around 20% of that amount might reasonably be expected as far as the handling of the regular lawsuit is concerned. As regards fees payable to the competent court, in the Court of First Instance the costs incurred by a patent action, including the filing of the lawsuit/pleadings/addendum and the attendance at the court on the hearing date, are estimated at around €373, while the relative cost of the procedure before the Court of Appeals amounts to €564. In addition to these sums, in actions for damages, the plaintiff must pay to the court a duty amounting to 6.5% of the total sum claimed by virtue of the lawsuit. Finally, application costs for injunctions include those for filing the application, drafting and submitting pleadings, and attending the court hearing; these will amount to €343.

As regards costs of recovery, the Greek judicial system operates a system where the loser pays the winner's costs, based on the discretion of the ruling court. However, such recovery is of a symbolic nature and is never a specific percentage of the actual costs suffered by the litigant parties. Moreover, in the event of reasonable doubt regarding the position of the defeated party at the outcome of the trial, the court may provide an exemption from payment of such costs.

11.   Hot topic(s)

In Greece, one of the most important recent developments in patent litigation was the inauguration of a special court department, which has existed and been functioning since May 2006 in the courts of Athens and Thessaloniki and which has exclusive competence to rule on patent-related cases. In this way, the need for judges specialising in intellectual and industrial property law issues has been met and the efficiency of the justice system in this field has been radically improved.

Additionally, there have recently been increasing numbers of lawsuits filed on behalf of companies and patent owners of branded products against pharmaceutical companies producing generic pharmaceutical products, claiming that their rights deriving from the patented and branded products have been infringed. The majority of such lawsuits for damages/injunction applications for the cessation of the infringement and prohibition of future infringements are currently pending, so the resulting case law is keenly anticipated in the legal world.

Ireland

John Whelan
A&L Goodbody

---

1. Lawyers

1.1 Representation of parties

The Irish legal system is similar to that in the United Kingdom in contentious matters and parties are represented by qualified solicitors and barristers, with solicitors preparing the case and barristers arguing it before the court at trial. Patent agents may also be called upon for advice in patent litigation before the courts, particularly where the case involves complex technology.

1.2 Choosing your legal representation

Although specialist representation is not required, it is advisable to use a solicitor and barrister with experience in patent disputes. Strictly speaking, solicitors have a right of audience in all Irish courts; however, it is usual also to instruct a barrister. It is necessary to instruct a local team of lawyers to manage a case; even so, English barristers have in the past been instructed to argue Irish cases before the Irish courts.

2. The court system

2.1 Overview of the civil court system

The Irish court system is divided into four courts of varying jurisdiction which are, from the lowest jurisdiction to the highest, the District Court, the Circuit Court, the High Court and the Supreme Court. The High Court deals with all civil claims where the value of the claim exceeds €38,092.14; however, in our experience all patent actions taken in the Irish courts have been taken in the High Court regardless of the value. The Supreme Court has appellant jurisdiction over the High Court's decisions.

An aggrieved party may only appeal a decision of the High Court on a point of law to the Supreme Court and there is no avenue for appeal on a question of fact, or direct access in patent cases to the Supreme Court. Typically, an appeal to the Supreme Court will take approximately 18 months to be heard.

2.2 The Commercial Court

While Ireland does not have a specialist patent court, patent infringement proceedings will generally be commenced in the Commercial List of the High Court, a relatively new development which is commonly referred to as the Commercial Court. The Irish Commercial Court has been conferred with specific jurisdiction over intellectual property disputes.

Specific permission from the Commercial Court has to be obtained before bringing the proceedings before it, which is done by way of an application for entry after the proceedings are issued. Entry into the Commercial List is presided over by Judge Peter Kelly, the President of the Commercial Court, and patent cases will be heard by either Judge Kelly or one of the other two Commercial Court Judges.

2.3 Background and experience of Irish judges

Whereas the judges in the general High Court may not have had any significant patent litigation experience, the judges of the Commercial Court have been involved in such cases. They do not necessarily have a technical background, however, and it is not unusual for an independent expert to be appointed to assist the court in patent disputes involving complex technology.

2.4 Decisions from other jurisdictions and the Irish courts

As there have been a limited number of reported patent decisions in Ireland, there is a likelihood that particular points of patent law may not have been considered by an Irish court. In the absence of any precedent on a particular issue, Irish courts will often look to the case law of other, particularly common-law, jurisdictions for guidance. Historically, Irish courts have demonstrated a preference for following the decisions of the English courts and such decisions are of persuasive authority in Ireland.

2.5 Patent-related actions in Ireland and requirements before actions are commenced

It is open to an alleged infringer to seek a declaration of non-infringement from the Irish court if it shows that it has applied in writing to the patentee for a written acknowledgement of non-infringement which was refused. There are, however, drawbacks to a declaratory judgment action in Ireland. Unless there are "exceptional circumstances" – a statutory term not yet defined by the courts – the party seeking the declaration must pay the costs of all parties.

A proactive defensive measure available to an alleged infringer is to petition the court to invalidate the relevant patent(s), if appropriate. Unlike the declaratory judgment action, the losing party would usually pay the costs of the successful party in the normal way, and the successful party could expect to recover most of its costs.

Where infringement proceedings are brought before the Irish court and the defendant counterclaims for invalidity of the patent, the infringement and invalidity actions will be dealt with simultaneously.

While there is no strict requirement to do so, it is customary to send a letter before action prior to commencing patent infringement proceedings. This letter will outline the nature of the claim and will typically require undertakings from the other party to cease infringement and deliver up or destroy infringing materials.

Failure to send such a letter before commencing proceedings may reflect badly on the plaintiff when the dispute first comes before the court, and if there is no case to answer this would have negative cost sanctions for the claimant. If the court deems that there is a case to answer, it may not be as amenable to the plaintiff's request to progress it at any speed if appropriate attempts by the plaintiff to deal with the dispute prior to proceedings had not been made.

3.   Procedure and timescale of proceedings

3.1   Typical timetable for a patent action

The Commercial Court has been widely acclaimed as a major success with regard to the speed at which it brings matters to trial. The system of rigorous case management which it has employed and its ability to offer early hearing dates has ensured the speedy resolution of disputes. The average time of a case from start to finish is about six to nine months. This is a marked improvement from the ordinary division of the High Court which can take as long as 18 months for a case to be heard. In either court, however, complex and contested patent cases are still likely to take longer than a typical commercial or intellectual property dispute.

A patent action will be commenced in the same way as any High Court dispute by issuing proceedings in the Central Office of the High Court. Either party can then apply to have the proceedings transferred to the Commercial List. The party making the application files a notice of motion, which is accompanied by a certificate of the applicant's solicitor outlining why it is appropriate that the proceedings be transferred to the Commercial Court. In our experience, almost all patent cases will be admitted to the Commercial Court, because they are of an intellectual property nature, which is one of the areas over which the Commercial Court expressly has jurisdiction.

Once admitted to the Commercial Court, the judge can make a number of directions at the pre-trial stage. These directions may set time limits for the exchange of pleadings, outline and clarify the issues in the proceedings, require discovery, or require the expert witnesses of both parties to consult with each other. The judge has significant powers to set time limits on the various stages of the proceedings in order to minimise costs, and the time limit on the exchange of pleadings is often set at two weeks. Timeframes are tight at all stages of proceedings in the Commercial Court and costs may be awarded against a party that fails to meet a deadline.

As the minimisation of cost and delay are core objectives of the Commercial Court, emphasis is put on opportunities for alternative dispute resolution, and the judge can adjourn proceedings for up to 28 days in order to allow the parties to enter into mediation.

The ability of the judge to get closely involved in case management is an innovative feature of the Commercial Court. A case management conference may be held with the parties to prepare for the case and for the judge to review whether pleadings have been sufficiently dealt with to date and whether his or her orders and directions have been satisfactorily complied with. The judge can set a timeframe for the conduct of the outstanding pre-trial procedures. A pre-trial conference is subsequently held. This establishes timeframes for the trial itself and, if the proceedings are ready to proceed to trial, a hearing date may be set.

3.2   **The potential effect of EPO opposition proceedings**
Where revocation proceedings are brought before the Irish courts, and opposition proceedings are pending before the European Patent Office (EPO) in relation to the same patent, there is a presumption – although some may say a relatively weak presumption – that the Irish proceedings will be stayed. However, the court will decline to grant a stay where convincing arguments can be made that it would be unjust to do so, particularly where the EPO proceedings are unlikely to be concluded for a significant time.

4.   **Availability of pre-action evidence gathering/disclosure**
There are general equitable rules adopted by the Irish courts, which allow a party to make an application for an *ex parte* search-and-seize order if it can be shown to the court that notice would cause the infringer to destroy potential evidence. Other than that, there are no specific rules regarding the gathering of pre-action evidence, and the ability to obtain disclosure or discovery of evidence in Irish patent cases usually only arises once proceedings have been commenced.

See below, at Section 6, for further detail.

5.   **Availability of interim relief (especially injunctions)**

5.1   **The criteria applied by the courts for the granting of injunctions in patent actions**
In the High Court, an application may be made for an interlocutory injunction restraining the defendant from any possible or actual act of infringement. The principles upon which an injunction will be granted are well established in Ireland and apply to patent cases in the same way they apply to other cases. In a case in 2003 (*Smithkline Beecham plc v Genthon BV* [2003] IEHC unreported, February 28 2003), the High Court held that the normal rules for interlocutory relief apply to applications for injunctions in patent infringement cases, but there have been no reported judgments on injunctions in patent cases since then. In the 2003 case, the plaintiff established that there was a serious issue for trial but failed to establish that damages would not be an adequate remedy, and the interlocutory injunction was refused.

In summary, the claimant must demonstrate that there is a serious question to be tried and that damages would not be an adequate remedy. The grant of an injunction is a discretionary power and the court will assess whether to exercise that power according to the balance of convenience. The plaintiff will be required to give an undertaking as to damages (to cover the situation where it is ultimately determined that it should not have obtained the injunction). Injunctions are more common in the High Court general, as judges in the Commercial Court may be more inclined to grant a speedy and early trial rather than interlocutory injunctive relief.

5.2   **Without-notice applications**
Interim injunctions may be granted on an *ex parte* basis (ie without notice to the other side), but applications for such relief will usually only be entertained in situations of urgency or emergency and will only be granted by the court for a short

period, such as a number of days. The injunction would last until the hearing of an *inter partes* interlocutory injunction application, at which both parties would be present.

5.3 Delay in bringing an application for an injunction

It will be necessary for an applicant for an injunction to demonstrate to the court that it has acted expeditiously. A delay in bringing such an application could prejudice its ability to obtain such relief, and injunctions have in the past been refused because the plaintiff did not move quickly enough.

5.4 Availability of protective briefs

A potential defendant does not have available to it the ability to file any document with the court or other body seeking notice to be provided should a claimant apply for *ex parte* relief. However, once *ex parte* relief is granted to a plaintiff, the court will require the applicant immediately to put the defendant on notice of the order and will have the parties come back before the court within a matter of days so that the defendant can present any defence it has.

5.5 Pros and cons of applying for injunctions

As with any proceedings, an injunction in a patent action, if granted, will be of significant benefit to the plaintiff, as it restricts the defendant from continuing to engage in infringing behaviour and preserves the status quo pending the trial.

The usual difficulties entailed in obtaining an injunction do, however, apply. In the absence of any Irish case law on the point (other than *Smithkline Beecham plc v Genthon BV*, discussed above), the Irish courts would be inclined to look to English case law for guidance. In England, interlocutory injunctions have recently been granted where it can be shown that severe price erosion will occur on the introduction of a generic drug.

It would be unlikely that any injunction application would give rise to a potential counterclaim for summary judgment, and even more unlikely that such a counterclaim would be successful, given the very high burden of proofs placed on an applicant for summary judgment.

6. Disclosure

6.1 When disclosure is available

Discovery is available in Irish proceedings, and generally arises once the exchange of pleadings between the parties has closed. Discovery may be carried out voluntarily by agreement between the parties or, if the parties fail to agree, by order of the court following an application by one or other party. In exceptional cases the parties can also agree to make discovery before pleadings have been exchanged, or can apply to the court seeking early discovery in proceedings that have been issued.

6.2 Extent of disclosure available

As a general rule, discovery will be required in respect of all documents relevant to matters in issue in the proceedings and which are necessary for the fair disposal of the proceedings or to save costs. Both parties are required to provide full disclosure of all documents relating to the dispute which are, or have been, in their possession, custody or power except those documents protected by privilege (which must be discovered, but not produced to the other side).

'Fishing expeditions' are not permitted and discovery can only be sought in relation to specific categories of documents specified by the party seeking them. That party must also give detailed and specific reasons as to why that category of discovery is being sought and should be discovered.

Given that patent infringement disputes will usually be dealt with by the Commercial Court, and that court encourages alternative dispute resolution, it is possible that limited discovery may take place at an earlier stage of the proceedings than is ordinarily the case. For example, the court may order that discovery of some documents be made prior to the parties entering mediation if one of the parties to the mediation does not have the same access to documents and would be at a disadvantage going into mediation.

Confidential legal advice will be privileged, as will documentation created for the purposes of enabling, or preparing, a party to defend anticipated legal proceedings.

6.3   Limitations on the use of documents disclosed

Under the rules relating to discovery, documents disclosed by one party to another party may only be used by that party for the purpose of the proceedings and cannot be used for any other purpose.

6.4   Protection of confidential information

There are no rules or procedures in relation to protective briefs under Irish law and there are no specific rules governing the disclosure of confidential information as part of any dispute, so it is usually left for the parties to agree, or seek directions from the court. It is, however, possible for the parties to enter into a confidentiality agreement which imposes restrictions in relation to documents discovered by either party in the proceedings; or at the request of either party the court may impose restrictions which are more or less similar to protective briefs in other jurisdictions.

7.   Evidence

7.1   How evidence is put before the court

The Commercial Court requires the exchange, prior to trial, of written statements based on agreed evidence, where there are issues in relation to which the respective parties' experts agree, but cross-examination of the witnesses at trial on these particular issues also follows. Separately, each expert will obviously be cross-examined at trial in relation to evidence that is in dispute between them (ie where the experts for each party disagree on a particular point).

The Commercial Court may also require the parties to set out their case in writing and use the witness statements, sworn under affidavit, as the main evidence of the witnesses. It is a feature of the Commercial Court to allow parties to exchange witness statements and expert reports pre-trial, aimed at reducing the length of proceedings.

Injunction hearings are heard entirely on affidavit evidence, and it would be rare to have any oral evidence.

7.2 Experts

Each of the parties to a patent dispute will instruct its own independent experts to give evidence. The parties will agree the number of experts, which will usually be one or two experts for either side depending on the complexity of the patents in dispute. As mentioned above, it is also open to the court to appoint its own separate independent expert to assist the court in complex patent disputes.

7.3 Cross-examination of witnesses

If a matter is pursued in the chancery division of the High Court, then the more traditional method of oral argument and cross-examination of witness at trial will happen. As mentioned above, the Commercial Court favours the exchange of written statements and the substitution of this for evidence-in-chief, but again cross-examination of witnesses at trial will always happen.

7.4 How the courts deal with experiments

There are no specific rules in Irish law or court procedure that deal with experiments in patent trials. Accordingly, data resulting from experiments will fall under the general law of evidence and the sitting judge has a wide discretion in how such evidence may be presented. There is a preference for oral testimony in the Irish courts at trial. Therefore, where a party proposes to use experiments, permission of the court will usually be sought in advance of the carrying out of such experiments, and the court will approve the mechanism by which the data and other results arising from the experiments will be admitted at trial.

8. Law

The principal statutory instruments governing patents in Ireland are the Patent Act 1992 as amended and the Patent Rules 1992 as amended.

The Patent Act 1992 implements the European Patent Convention (EPC) in Ireland and follows closely the wording of the EPC and the Community Patent Convention. The Patent Rules deal with the administration of patents by the Patents Office and with the powers of the Controller of Patents.

8.1 The test for patent infringement

The language of the Irish legislative provisions which deal with direct and indirect infringement is almost identical to that contained in Articles 25 and 26 of the Community Patent Convention. In brief, a patent is infringed if a third party makes, offers for sale, puts on the market, imports, or uses a product or process which is the subject matter protected by the patent, without the patent owner's consent.

The Irish courts have endorsed the purposive approach to the construction of patent claims in line with the approach under the Protocol on the Interpretation of

Article 69 of the European Patent Convention, to which Ireland is a contracting state. The purposive approach means that patent claims will be construed by the Irish courts objectively through the eyes of the skilled addressee. The Irish courts have not adopted a doctrine of equivalence in patent proceedings, and the doctrine does not as such form part of Irish patent law.

8.2   Assessment of patentability

The provisions dealing with patentability reflect Articles 52 to 57 of the European Patent Convention and provide that an invention is patentable if it is susceptible to industrial application, is new and involves an inventive step. Something is considered to be new if it does not form part of the state of the art, which includes everything made available to the public (whether in Ireland or elsewhere) by means of a written or oral description, by use, or in any other way before the date of filing, or priority date if different, of the patent application.

The presence of an inventive step depends on whether the invention claimed would be obvious to a person skilled in the art, at the filing or priority date of the patent application.

Insufficiency is dealt with in the same manner in Ireland as in the United Kingdom and an application for revocation of a patent may be made on the grounds that the specification of the patent does not disclose the invention in a manner sufficiently clear and complete for it to be carried out by a person skilled in the art.

8.3   Patentability assessment of certain subject matters

The following are not considered to be inventions which are patentable under Irish law:

- a discovery, a scientific theory or a mathematical method;
- an aesthetic creation;
- a scheme, rule or method for performing a mental act, playing a game or doing business, or a program for a computer;
- the presentation of information;
- a method for the treatment of the human or animal body by surgery or therapy;
- a diagnostic method practised on the human or animal body;
- a plant or animal variety or an essentially biological process for the production of plants or animals other than a microbiological process or the products thereof; and
- an invention the publication or exploitation of which would be contrary to public order or morality.

As with most European countries, software programs are generally not patentable in Ireland and are, instead, protected by copyright as a literary work; however, it is possible to obtain software-related patents.

The EU Biotech Directive (98/44/EC) has been implemented in Ireland by Regulations which set out a non-exclusive list of subject matter which is patentable and non-patentable. Some biotechnological inventions are not patentable on the grounds of public order or morality, such as processes for cloning human beings, for modifying the germ line genetic identity of human beings, the use of human embryos for industrial or commercial purposes, and a process for modifying the genetic identity of animals which is likely to cause them suffering without resulting in any substantial medical benefit to man or animal.

9. Available remedies

The remedies typically available to a patent owner include: order for delivery up or destruction of infringing goods, an injunction preventing further infringement, damages or an account of profits, and costs. Punitive damages are generally not available, and have never been awarded, in a patent infringement case. Furthermore, damages should not be awarded against an "innocent infringer" who was genuinely not aware, nor should have been aware, of the existence of the patent.

9.1 The Irish Transposition of Enforcement Directive 2004/48/EC

Ireland transposed the Enforcement Directive into national law in July 2006. It is too early to tell what impact this will have on Irish law. That being said, many of the measures provided for in the Enforcement Directive were already present in Irish law. Other mandatory provisions of the Enforcement Directive, which were implemented into Irish law, introduced new options for intellectual property holders primarily aimed at tackling counterfeiting. A claimant now has the ability to seek a court order for disclosure of information from someone allegedly involved in infringing, so that the claimant can establish the chain of parties involved in the infringement.

9.2 Calculation of financial remedies

A claimant may seek damages or an account of profits, but cannot be awarded both. However, the claimant can seek disclosure of, for example, the financial records of the infringer in order to determine which financial remedy will be sought. The principles applied by the courts in determining damages are similar to those applied by the English courts. Once infringement has been established, the courts will seek to compensate the claimant but not to punish the infringer. Accordingly, damages will be calculated on the basis of the loss made by the claimant, which may result in a calculation based on the loss of profits suffered by the claimant. If such a calculation is not possible, the court may award damages on the basis of the royalties which the infringer would have paid to the claimant if a licence had been granted. This will be a reasonable royalty, calculated on the basis of the market rate and not what the either of the parties to the proceedings propose. The calculation of the royalty rate will be a matter for the court to decide on the basis of the market evidence put before it.

9.3 The effect of an appeal on enforcement

A successful claimant in a patent action should bear in mind that its ability to enforce any order of the High Court is likely to be stayed pending an appeal (if there is one) to the Supreme Court.

10.   Costs

Litigants in a patent action before the Irish Commercial Court should expect to pay in excess of €150,000 for legal costs, that amount increasing depending on the complexity of the case. As in the United Kingdom, the losing party must generally pay the successful party's legal costs. A successful litigant should expect to recover in the region of 70 to 75% of the actual costs incurred.

11.   Hot topic(s)

In terms of legislative developments, the Patents (Amendment) Act 2006 brought Irish patent law fully into line with obligations prescribed by TRIPS (the international Agreement on Trade-Related Aspects of Intellectual Property Rights), EPC 2000 and the Patent Law Treaty, while also reflecting clarifications which had been provided in recent case law in the intellectual property area and streamlining certain administrative procedures before the Irish Patents Office.

Among other changes to Irish law, the 2006 Act aims to clarify the grounds upon which patent revocation is available. Previously revocation could be claimed where "the protection conferred by the patent has been extended by an amendment of the application or the specification of the patent". This contrasted with the simplicity of the equivalent ground under the EPC, that is, that "the protection conferred by the patent has been extended". The Act attempts to bring the Irish provision into line with that of the EPC, and now provides that revocation can be claimed where "the protection conferred by the patent has been extended by an amendment which should not have been allowed".

Prior to the Act, the law contained a groundless threats provision. However, whilst the availability of the relief was intended only for those unjustifiably threatened with 'indirect' patent infringement, the wording of the relevant section could catch patentees who threatened direct infringers (ie those responsible for manufacturing or importing a patented product or using a patented process). The provision has now been amended to confirm that the groundless threats action will not be available to direct infringers where they are accused not only of the relevant acts of direct infringement but also of indirectly infringing the patent (eg by selling infringing goods).

The Commercial Court has brought a new regime for the resolution of intellectual property, and in turn patent, disputes. Prior to this, patent litigation was not particularly common in Ireland, as cases would often be left languishing in court lists with reluctant parties adopting delay tactics that severely hampered dispute resolution. However, patent proceedings are now being actively case-managed, pleadings are much more precise, interrogatories are encouraged instead of lengthy discovery requests, and witness statements and experts' reports are required to be exchanged prior to the hearing, all resulting in the parties being better prepared, and the courts being more willing, for early trial. The Commercial Court experience has been a wholly positive one and patent cases that historically might have taken two to three years to get to full trial are now in some cases being disposed of in a matter of months.

The judges appointed to the Commercial Court have a keen interest in intellectual property, and this has been key to its success in this area. In relation to interlocutory applications, a practice has developed whereby costs have been awarded at the interlocutory stage – a very powerful mechanism in the fight against intellectual property infringers. That said, in patent cases, the judge will often favour an early trial over interlocutory relief. Given the speed with which cases are now dealt with in the Commercial Court, and the fact that intellectual property disputes are often multi-jurisdictional, this has resulted in clients being more confident about the prospect of Ireland as a jurisdiction in which to litigate, based on the likely completion dates for trial and the opportunities to have cases dealt with on a preliminary issue basis.

Ireland summary
- As in the United Kingdom, both solicitors and barristers represent parties to a patent claim.
- Patent trials will generally be held in the Commercial Court, which can process a claim from beginning to end in six to nine months. Complex cases may take longer.
- Ireland does not have a specialist patent court, but as most claims will be dealt with by the Commercial Court, the judge hearing the case is likely to have experience in dealing with patent disputes.
- Patent trials will usually cost at least €150,000 and the 'loser pays the winner's costs' system applies in Ireland.
- The Irish High Court will treat English patent decisions as persuasive authority, and an appeal from the High Court can only be made to the Supreme Court on a point of law, the latter being 'the end of the road' for patent litigants.

Italy

Giovanni Francesco Casucci
Casucci Law Firm

1. Lawyers

Court proceedings in Italy regarding patent cases are characterised by the absence of a technical judge. In particular, any patent case discussed before the civil court usually involves a court expert appointed by the judge, in order to support him in ruling on validity/infringement issues. Accordingly, litigation teams will generally include a patent attorney. The team selection is usually organised by the client's preferred partner (sometimes the client's lawyer, sometimes the client's patent attorney).

Only lawyers are admitted to plead the case before the court. Nevertheless, the patent attorney plays a very important role in proceedings, as he is the natural counterweight to the court expert in assessing validity and infringement issues. In practice, when the judge opens within the trial the court expert opinion phase (*Consulenza Tecnica d'Ufficio* – CTU), the court expert witness has 90 days (and sometimes longer) to write and file his opinion and within this period the judge grants the parties' expert specific deadlines to deliver his technical reports. (Usually two or three reports can be exchanged between them and the expert witness.) The court expert can organise technical hearings, inspections and so on.

There is no territorial limitation on lawyers and they can plead cases before any competent Italian court. Nevertheless, it is usual practice to appoint a local lawyer if the litigation court is different from that of the main lawyer. This is because a local lawyer:

- will know the local practice of the office of the court's clerk and might therefore more quickly be able to obtain information about the case and/or copies of useful documents; and
- will be familiar with the habits and personalities of local judges and will be better able to indicate the specific attitudes of judges involved in intellectual property (IP) cases.

A power of attorney is needed to appoint a lawyer. Powers of attorney must be in writing and, if relating to a particular litigation case, must be physically attached to the writ of summons. The power of attorney may be valid for both first instance and appeal phases. For the Supreme Court, it is necessary to sign a new special power of attorney. If a person is a citizen of, or a company has a registered office in, a member country of the Hague Convention of October 5 1961 and wishes to act in Italy, the power of attorney must include an apostil under the procedure set out in that convention. The proxy must also be legalised and notarised according to the national provisions of the relevant country.[1]

Despite the existence of specialised IP civil courts, there is no official list of specialised IP lawyers (as in France, for instance). The only way to identify such experts is to use AIPPI and LES guides, or the selections made by Chambers and Partners and Legal 500.

2. The court system

Italy is a typical civil law system. There are 165 first-instance courts (*Tribunali*) within the Italian territory, 29 courts of Appeal (*Corti d'Appello*), based in certain districts and one Supreme Court (*Suprema Corte di Cassazione*), based in Rome.

As an exception to the general system, the court system in patent cases (and in general for all contentious IP matters including unfair competition[2]) is now characterised by the exclusive competence of specialised IP court divisions (*sezioni*). On July 1 2003,[3] 12 courts were appointed to deal with all national IP cases, according to territorial competence.

The Italian civil IP specialised court's divisions (hereafter referred to as IP courts) are exclusively competent to deal with any actions relating to patents, including matters of validity and infringement. No other jurisdiction is entitled to deal with these issues.

The workload of the IP courts during the last five years shows that only a few of them deal with significant numbers of cases. The most active, and therefore experienced, courts are those in Milan, Venice, Rome, Turin, Naples and Bologna. The rest do not have heavy workloads.

The appeal procedure in IP cases is also consistent with the exclusive competence of specialised court departments located in the same 12 towns. The appeal structure consists of a panel of three judges who review the decisions rendered by the court of first instance. Both questions of law and of fact can be dealt with anew by the territorially competent court of appeal (*Corte d'Appello*). While it is not possible to change claims (*domande*) filed at first instance, it is possible to ask for a proceeding phase to examine evidence which was not fully examined at first instance. The average length of appeal proceedings is about two years.

The Supreme Court (*Corte di Cassazione*) is located in Rome and its aim is to review decisions rendered by the lower courts, exclusively on matters of law. Decisions of appeal courts can be overturned due to an incorrect interpretation of the

---

1   According to article 4(4) of EC Regulation 1348/2000, the apostil is not necessary for countries which are part of the European Community, except for Denmark.

2   The issues involved relate to cases of know-how, unregistered trade marks, slavish imitation and unregistered design rights.

3   Under Decree no 168 of June 27 2003, 12 courts in the following towns were designated: Bari, Bologna, Catania, Florence, Genoa, Milan, Naples, Palermo, Rome, Turin, Trieste, and Venice. These designations are preliminary according to Section 16(4), and within two years from the entry in force of the Executive Decree the government is entitled to review the number and location of these specialised courts. Therefore, in principle, a modification of the number or location of the courts would be possible. Even now, it is obvious that Sardinia is underrepresented while there is duplication in Sicily (Palermo and Catania). Moreover, the major area for border controls (Gioia Tauro, near Reggio Calabria) falls under the jurisdiction of Catania for civil enforcement. Therefore, a modification is strongly recommended.

law, and will thus be sent back to the lower court. If the question does not require a new evaluation of the evidence or interpretation of the facts, the Supreme Court will decide the matter directly without referring the case back. This court is also competent to decide on conflicts concerning jurisdiction, or issues of competence arising in lower courts. Decisions of the Supreme Court are final and can be reviewed only in exceptional cases. The average length of Supreme Court proceedings is about two years.

Only IP courts are competent to deal with the main patent issues (ie validity, revocation and infringement). The Italian Patent Office is only entitled to manage the administrative steps relating to compulsory licence (which is very rarely claimed).

When choosing the appropriate competent court, there are three criteria to be considered:

- the forum of the defendant's domicile (ie where the defendant has its registered office);
- the forum of the harmful event (any place where the infringement has occurred); and
- the forum of the patent's elected domicile (only for direct nullity actions).

Thus, according to specific expectations relating to the workload of the IP courts, or to their approaches and specific experience, it is possible to use the above-mentioned criteria in order to claim jurisdiction in the most favourable IP court.

In cases of non-infringement declaratory actions (NIDA), the forum frequently invoked is the forum of the claimant. The reason for this is based on a special interpretation of the concept of 'possible harmful event'. According to this reasoning, the possible prejudice (ie harmful event) to be suffered in cases of an infringement action wrongfully based would be the main premises of the alleged (innocent) infringer. It is worth noting that this kind of action (known as a 'torpedo' action) is based on the rule of *lis pendens*, according to the Lugano and former Brussels Conventions.[4] The practical advantage deriving from such actions is to try to freeze enforcement actions in 'speedy' countries by moving first with a defensive action (mainly of non-infringement) in a 'slow' country. Accordingly, in this case, the second action (by the rights owner) in principle has to be suspended until the (slow) decision of the first seized court is issued.[5]

Judgments issued in other jurisdictions may have some influence in Italian IP courts. For example, recently reported judgments in German and UK proceedings (particularly in terms of Lord Hoffman's observations regarding equivalency (the

---

4   Article 21 provides that when the same action is pending before a court of a member state, the court of another member state subsequently seized must, on its own motion (*ex officio*), decline jurisdiction. The case should be pending between the same parties and have the same object and the same cause of action. Article 22 says that when the two cases are not identical but related, the second court may suspend the proceeding. EC Regulation 44/2001 has implemented the provisions contained in the Brussels Convention which are now binding on all member states with some slight amendments.

5   The term 'torpedo' was selected in order to express the peculiar concept of the speediness of this strategy: "It is a well-known principle of sea warfare that an escorted convoy should travel at the speed of the slowest ship. If the various ships travel at their own speed, there will be no convoy and no unity."

Catnic Test)) have been accepted and applied. Such influence is strongly supported by various initiatives (promoted by the EPO Academy and the WIPO Academy) which are designed to foster exchanges of experience and ideas among IP specialised judges and offering various events for common discussion.

In Italy, there is no mandatory rule for sending warning letters before commencing civil actions. (This contrasts with the position in other jurisdictions, for example in relation to damages claims in Germany.) In the past, a warning letter was a fairly common means of managing pre-trial conflicts. Following the inception in 1997 of the torpedo practice mentioned above, however, the receipt of a warning letter was considered sufficient reason for opening a NIDA with the aim of trying to paralyse the attack. Therefore, from that time on, the use of warning letters has been strongly discouraged in Italy.

3. Procedure and timescale of proceedings

It should be noted that, in order to commence litigation (via ordinary or urgent proceedings) in Italy, it is not necessary that a patent should have been granted. It is sufficient if the application is deemed public (even if in Italy there is no official publication made by the Italian Patent Office) or simply notified to the counterpart (article 132 Italian IP code). Thus, with regard to European patent applications, it will be sufficient to file a translation of the application in Italian before the Italian Patent Office (UIBM). This grants a strong competitive advantage compared with other countries (eg Germany), where litigation can be initiated only after the grant of the patent right.

Accordingly, a pending EPO opposition would not necessarily have the effect of holding up the procedure. On the merits of the given case, the judge may deem it appropriate to stay the proceeding, but this is unusual (or even exceptional) at the preliminary measures stage.

In terms of procedure, the ordinary Civil Process Code is currently applicable, following decision no 170 of May 17 2007 of the Constitutional Court (*Corte Costituzionale*) which declared inapplicable to IP litigation the special procedural rules applied for company law conflicts (which had been temporarily applied for the previous three years).

Civil procedures in Italy are based on written claims rather than oral arguments. Therefore, the main work of the attorney is to draft written briefs. Nevertheless, in preliminary proceedings oral hearings usually play a decisive role in presenting the parties' arguments before the judge.

In order to understand patent litigation in Italy in relation to time issues, it is important to note that there are two different procedural means of enforcing IP rights:

- ordinary proceedings upon the merits; and
- urgent proceedings.

3.1 Ordinary proceedings upon the merits

Ordinary proceedings upon the merits are the same as those for infringement declaratory actions and for nullity actions against a patent. The typical steps are set out below.

(a)   *Serving the writ of summons*

The ordinary action normally starts with the service and filing of the writ of summons (*atto di citazione*) requesting the defendant to appear at a given hearing, no sooner than 90 days from the date of service for a national defendant and 150 days for a foreign one.

The writ of summons should contain all the facts, legal arguments and evidence to be used during the trial. Finally, the writ must contain a precise claim. The writ of summons and claim delimit the subject matter of the litigation. The claim cannot be extended and neither is it possible to introduce new arguments or means of evidence during the trial.[6] The defendant must file his statement of defence at least 20 days before the date of the hearing if he makes a counterclaim.[7]

(b)   *First hearing*

At the first hearing, the judge will verify the regular service of the writ of summons and the timely filing of the counterclaim (if any). At this hearing the parties may more precisely define their claims or request that other parties be summoned. The claimant may also add new claims that arise as a consequence of the counterclaims of the defendant. If, as is usually the case, a request is made to the judge, he may fix three kinds of term in order to allow the parties to:

- formalise their respective claims (30 days);
- reply to opposing arguments and argue about the evidence to be added to the proceeding (ie documentary evidence or evidence which may arise from laboratory tests or experiments) (30 days); or
- reply in order to argue about counter-evidence (20 days).

The judge will fix a specific hearing in order to discuss such issues (art 184 of the Italian Civil Procedural Code). If requested by the parties, the judge may fix a specific hearing in order to allow the appearance of the parties and to attempt to settle the case.

(c)   *Evidence hearing*

Following the exchange of briefs as outlined above, at the evidence hearing the judge examines the various forms of evidence and admits or rejects them (totally or partially), including the results of the search order proceeding. The most important evidence in patent cases comprises:

- the court expert witness phase for infringement/validity issues;
- the court expert witness for the damages assessment; or
- the formal interrogatory of the counterpart legal representative, in order to define the sales turnover or other issues related to the damage identification.

---

6   Section 184 of the Code of Civil Procedure allows a term before trial for the claimant to review his claim or requests for evidence in view of the defendant's statements of evidence. After this opportunity, no more changes will be allowed. Nevertheless, a party not complying with the time limits can ask the judge for an extension in exceptional circumstances.

7   This deadline only applies to counterclaims of the defendant, not for ordinary defences asking for the action to be discussed.

The measure under the first bullet point above is usually granted by the judge. The other two are granted only if the outcome of the first point leads to a conclusion of infringement or a confirmation of validity. These measures will therefore come at a later stage in proceedings.

(d) Court expert appointment hearing and the court expert phase management

The judge usually appoints a patent attorney as court expert. The parties may agree to select a list of possible names (so as to avoid conflicts). The judge, however, is free to appoint any expert, regardless of the parties' recommendation.

The judge appoints the court expert and fixes a subsequent hearing in order to define the questions to be investigated by him. At this second hearing, it is very important for the parties to discuss these questions in order to focus the court expert's attention on specific issues.

The judge usually grants the court expert a term of 90 days to deliver his opinion (unless specific reasons may justify a longer term) and fixes the hearing for the discussion of the expert's findings. In order to save time, it is common practice for the judge to require the court expert to exchange his preliminary opinion with the party experts and receive their comments and finally to file the definitive opinion. In this way, the subsequent hearing will be devoted to discussing the juridical consequences of the factual opinion provided by the court expert. Each party will appoint a party expert, entitled to discuss the technical issues with the court expert.

The court expert will usually assign the party experts two court delays in order to deliver their technical briefs about the questions indicated by the judge. Practically, a third round of briefs is usual in order to finalise the arguments raised by the party experts. During this phase any party is allowed to file new evidence (with the exception of terms indicated above). Therefore, the alleged infringer usually waits for this phase in order to provide new prior art, thereby limiting the reaction time of the patentee. In order to assess the infringement, a party expert may require the court expert to deal with experiments and technical inspections.

(e) Hearing discussion of the court expert opinion

At this hearing, the party's lawyers discuss the opinion of the court expert and finalise any request for further investigation regarding damages (if the results are likely to convince the judge as to the existence of an infringement). If the judge finds that he has sufficient information and that there is no need for further investigations (ie damages can be assessed on the basis of the evidence already provided, or in an equitable way), then he will also decide to fix the final claim hearing. If it looks likely that one of the parties will be able to contest the court expert phase, the judge may decide to renew the phase appointing a new court expert (this is very rare), or to fix a discussion hearing with the court expert in order to discuss the case and ask him to clarify the doubts raised by the parties.

(f) Final claims hearing

Despite the fact that all the previous phases are generally executed within a timeframe of about a year and a half (or sometimes less), the final claims hearing is usually fixed very late (ie usually ten months later). This is because, until now, the IP courts have not dealt exclusively with IP cases but also with other civil issues, and this has inevitably had an impact on their workload.

At the hearing, the parties simply summarise their claims according to the results of the previous instruction phase. The judge then declares the case ready for the decision before the judges' panel (*Collegio*), assigning the parties a term of 60 days for their conclusion briefs and 20 days for conclusion replies. The final decision is usually taken without a final discussion hearing, unless the parties request it. In this case, a hearing date will be assigned and the judges' panel will hear the case before taking the decision.

3.2 Urgent proceedings

Urgent proceedings are necessary in order to obtain a preliminary measure (eg seizure and/or injunction). The typical steps are set out next.

(a) Filing the application

The procedure for a preliminary measure (eg a search order, a preliminary injunction, or seizure) starts with the filing of an application before the office of the court's clerk of the competent IP court. The application must contain:

- a clear indication of the measure requested;
- the factual and juridical grounds underlying the request; and
- an indication of the future actions to be taken.

The proceeding is pending from the filing date and the docket number is assigned.

The file is then assigned by the presiding judge to an instructing judge, who will take the appropriate measures. The only exception relates to description orders, which must be granted exclusively by the presiding judge.

The requested measure may be granted *ex parte* or *inter partes*.

With the exception of search orders (which do not involve any other judicial activity or confirmatory hearings), the preliminary seizure and/or injunction order, if granted *ex parte*, will usually provide for the fixing of a hearing to discuss the case after the service and/or the execution of the order.

If granted *ex parte*, the order must be executed within a time limit of 30 days from issue and, within this timeframe, the applicant must also file the action upon the merit (complying with the requirements of TRIPS and Article 9.5 of the EU Enforcement Directive).[8]

(b) First hearing for the case discussion

At this hearing the judge will discuss the case before the parties. The proceedings are

---

8   Article 131 *bis* of the Italian IP Code. This article contains a misleading indication (para 1-*quater*) regarding the fact that a preliminary injunction measure does not require the opening of an ordinary proceeding if "capable of anticipating the final decision", thus treating a preliminary decision as a definitive judgment. Such indication should be disregarded in light of the clear provisions of the TRIPS Agreement on Trade-Related Aspects of Intellectual Property Rights (Article 50.6) and the Enforcement Directive (Article 9.5). An amendment is strongly suggested and it is likely that a judge will raise an issue before the Constitutional Court.

informal and the parties will be free to introduce arguments, documents and even witnesses. In patent hearings, the judge will immediately consider appointing a court expert in order (summarily) to evaluate the likelihood of validity and infringement. The judge, appointing the court expert, usually grants him a shorter timeframe (ie usually around 30 days) within which to present his opinion. Nevertheless, the court expert usually requests (limited) postponements.

The appointment of the court expert and the following phase is the same as the process outlined above for ordinary proceedings. Following the appointment, at the same hearing, the judge will fix the hearing date for discussion of the court expert's opinion and the final discussion between the parties.

(c) *Second hearing for the case discussion*

At this hearing the parties will discuss the whole case, including the findings of the court expert, and the judge will finally will close the discussion in order to come to a later decision regarding the requested measure. The decision is usually filed by the office of the court's clerk and communicated to the parties within 15 days.

(d) *Appeal against the decision of the first judge*

It is possible to appeal the decision of the first judge (*Reclamo*) within 15 days from the date of communication or service. The appeal is filed, as the application at first level, before the office of the court's clerk, and a hearing date is usually fixed in order to discuss the case before a judges' panel of the same IP court.

The parties will have only one chance to discuss the case before the judges' panel, and after the hearing a decision will usually be given within 15 days (sometimes later).

There is no additional appeal after this hearing. The judge in the main proceeding may take note of any variation to an order given or refused in the appeal.

4, Availability of pre-action evidence gathering/disclosure

In Italy there is a specific pre-trial means of evidence collection (ie the search order – *descrizione*).[9]

This order (available in Italy since before 1979) used to be quite rare in the European Union. Only France has an analogous proceeding (*saisie-description*). The *descrizione* is an order granted by the president of the competent court authorising the plaintiff (assisted by the bailiff, eventually by a court expert and usually with a photographer) to inspect and describe the allegedly infringing products/processes. Italian case law has recently begun to interpret the concept of 'evidence of the infringement' more broadly, extending the power of the bailiff even to collection of documentary and accounting evidence relating to the business involved in the alleged infringement. This approach is particularly useful and significant in enforcement strategies, because it offers the opportunity to collect *ex parte* all the elements to quantify the damages suffered, without the risk of manipulation that is typical during the main proceedings. The approach is also perfectly in line with the

---

9   Article 128 of the Italian IP Code.

policy envisaged by EC Directive 48/2004, Section 7, which *de facto* represents a codification of the Italian legal practice.[10]

Although currently there are no officially decided cases, it could be argued that there is a possibility of obtaining such an order in Italy, so as to execute it in other EC countries by virtue of EC Regulation 1206/01 of May 28 2001 on cooperation between the courts of the member states in taking evidence in civil or commercial matters. According to this regulation, it is possible for a court of one member state to request the competent court of another member state to collect evidence.[11] The requesting court may ask for the execution in accordance with a special procedure provided by the law of the court's own country (ie a *descrizione* order), and if provided by the law of that country, the parties, and their representatives, if any, have the right to be present at the collection of evidence by the requested court.

An Italian order issued by the Court of Genoa on May 5 2005 against an Italian reseller and a UK-based company can be seen as a precedent. The UK authorities refused to execute the order, so the Court of Genoa raised the case before the European Court of Justice. The case (C–175/06, TEDESCO/RWO) was heard by a plenary hearing, and seven European member states participated in the case, presenting their position with the European Commission. The Advocate General Juliane Kokott filed his conclusion on July 18 2007, confirming the fact that the Italian *descrizione* is an evidence-collection tool which can be executed throughout the European Union, thanks to EC regulation 1206/01. Unfortunately, the case was settled and the European Court of Justice decided not to prosecute the case. Nevertheless, the ECJ advocate general confirmed the approach of the Italian case.

The oral proceedings will contain a detailed description of the operation and the evidence collected. The purpose of this procedure is to collect official evidence of infringement for the purpose of using it in the trial. Normally, the order can be obtained ex parte (*inaudita altera parte*) if there is a risk that the defendant may remove all evidence of the allegedly infringing goods. Usually, no bond is requested. Requests for a search are frequently made during expositions or trade fairs because no civil seizure is allowed in these places. After the order for a search is issued, the plaintiff has 30 days to summon the defendant for an ordinary trial. If this deadline is missed, all the evidence collected will be invalid and cannot be used for any future procedure.

The collected evidence must be officially authorised by the judge in order to be admitted into the main proceedings. After this (as indicated above), this evidence can be used in any other context (ie parallel proceedings abroad).

---

10   Despite the better reputation of the French legal practice, the EC directive seems to the author to be more inspired by, and consistent with, the Italian legal practice. Evidence of this can be found in the possibility (implied in the EC directive) of the participation of the patentee and his consultants and lawyers in the search operations. This possibility was officially denied by the French institute before implementation of the EC directive.

11   This specific strategy (called *Italian Intruder*) of using the *descrizione* and the EC Regulation 1206/01 as a strategic tool in IP enforcement was announced by the author at the CEIPI 40th annual conference in Strasbourg, April 23 2004, and will be the subject of an article to be published shortly.

5. Availability of interim relief

The Italian IP code expressly provides for injunctions (*inibitoria*) as interim relief.[12]

An injunction is an order to cease and desist from using a product or process which allegedly infringes the patent by manufacture, promotion, distribution and so on. Following implementation of the TRIPS Agreement, it is now also possible to obtain an injunction on a preliminary basis.

The order is subject to two requirements:

* urgency; and
* the likelihood of the facts and legal issues underlying the reason given for the application by the claimant.

The first implies that the claimant will suffer irreparable harm as a result of delays to the ordinary procedure. Moreover, the claimant must be able to prove that he has reacted speedily on becoming aware of the infringing actions. A delay of more than six months on becoming aware of the alleged infringement may cause the rejection of the application for interim relief.

The second point above implies that evidence of the infringement can be reasonably easily assessed on a preliminary basis. In the patent field, such evidence may be difficult to assess on a summary basis. In any event, judges usually decide to grant preliminary measures after the appointment of a court expert, who has the task of assessing the likelihood of the infringement. The expert's report should be limited to a summary analysis of validity/infringement issues.

The order may be granted *ex parte*, but in practice the IP courts very rarely grant *ex parte* injunctions. The judge will usually grant a hearing in order to discuss preliminary issues and then appoints a court expert witness (following the procedure outlined earlier).

The court can also be requested to impose a penalty (*astraintes*, namely monetary fines) in cases of delay in complying with the order. Contempt of an injunction granted by the court also constitutes a criminal offence.[13]

Protective briefs, as known and largely applied in Germany, are practically unknown in Italy. It might sometimes be useful to file before the presiding judge of the specialised IP court a request not to not grant *ex parte* injunctions or description orders during expos/trade fairs. The Court of Milan always denies the validity of such requests, but other IP courts are more ambivalent.

In terms of judging pros and cons, a request for an injunction does not necessarily have any particularly negative effect. Any judicial move against an alleged infringer will usually result in a counterclaim for nullity of the patent involved. Perhaps the major judgement to be made in terms of pros and cons is whether to apply for the preliminary injunction *before* the opening of the ordinary action or *within* the ordinary pending proceeding.

In the author's opinion, the filing of an injunction request before the ordinary proceedings may offer the other party the chance to avoid the pressure of litigation, if the

---

12   Article 131 of the Italian IP Code.
13   Article 388 of the Criminal Code: "Malicious failure to execute a judicial order".

request is rejected, and oblige the patentee to institute new procedure upon the merit.

On the other hand, the filing of a preliminary injunction request during the ordinary proceedings, even if rejected, may help the claimant to understand the position of the judge (who will also be the judge for the final decision). In fact this may enable a change in strategy during the full trial, providing new evidence in order to counteract a previous opinion. At the same time, the alleged infringer will be faced by the continuous pressure of the litigation.

6.   Disclosure

The so-called disclosure order was implemented in Italy just after the coming into force of the TRIPS Agreements (through article 77 of the Patent Law (1996)). Early court experience of such orders was very limited. The first limitation related to the non-enforceability of such orders. The party requested by the judge to give disclosure was not obliged to do so, and could simply state his intention not to disclose. Therefore, this procedural means was considered not to be very useful.

The best means of obtaining forced disclosure is the description order, the aim of which is also to collect any possible spontaneous declaration of the defendant whilst the execution takes place. During the description operations the defendant is invited to declare his position regarding the legal and factual issues relating to the alleged infringement, so there exists a possibility that he may admit liability or divulge relevant evidence.

Following the entry into force of the Enforcement Directive, the Italian IP Code (amended in 2006) has been duly amended through a new version of the disclosure order, called *Diritto d'Informazione* (article 121 *bis*). According to this, a judge, at the request of a party, may order any person involved to disclose any information relating to the infringement chain and connections. A party who refuses to comply with the order, or who supplies incomplete information, is regarded as being guilty of the crime of "fake deposition" (article 372 of the Italian criminal code).

Although this new approach may offer more opportunities to obtain better results, the only situation where such an order has taken place is when the court, despite initiating a court expert phase for damage assessment, decides to obtain disclosure relating to profits realised by the infringer in order to calculate damages.

Once any evidence and/or declarations become part of the process docket, such evidence may also be used in other procedures, unless the judge decides to keep such evidence confidential at the request of the disclosing party.

7.   Evidence

The evidence must usually be submitted to court within the specific time limits indicated.

Documentary evidence must be submitted within the specific timeframe and briefs to that exclusively deputed. Nevertheless, according to article 121,[14] during the

---

14   Article 121 (Allocation of the burden of proof): "... 5 *In matters set forth in this code, a court-appointed expert can receive documents regarding the questions posed by the judge, even though they have not been filed in the proceedings, by disclosing them to all parties. Each party can appoint more than one expert.*"

court expert phase any of the parties' consultants may add new evidence before the court expert.

As regards non-documentary evidence (eg declarations, witnesses, oaths), the request must be submitted before the court within the requisite time limits, and the court will decide on their admissibility at the evidence hearing.

Witness hearings (which are rare in patent cases) are carried out by the judge. The opposing party is allowed to cross-examine witnesses (and there is a very limited power to introduce issues which are different from those which were originally raised). Court experts are appointed as described above.

Possible experiments must be requested officially by the court expert and authorised by the judge. They may be carried out in laboratories or during on-site visits, but always under the control of the court experts, with the participation of the party experts.

8. Law

The law relating to patents (ie in terms of substantive and enforcement issues) is entirely constituted by the Industrial Property Code, which came into force in 2005 and was subsequently amended in light of implementation of the Enforcement Directive (48/2004).

The equivalence test is applied according to the Catnic rules, and claim interpretation is usually assessed according to the protocol of EPC article 69.

The validity requirements will normally be evaluated by the judge with the support of a court expert, who will usually be a patent attorney with a European qualification.

Judges rarely disagree with the position adopted by a court expert. Nevertheless, in most experienced courts (eg Milan) judges will frequently choose not to follow the position of the court expert if there is some misinterpretation of the legal concepts underpinning novelty (the publicly accessible nature of the prior teaching) or the scope of protection.

9. Available remedies

The remedies available are:

- definitive injunction;
- seizure/recall of goods;
- destruction of goods;
- damages; and
- publication of the decision.

These measures[15] are fully compliant with the Enforcement Directive.

Damages are usually assessed within the same proceedings, except where the parties may agree to see the decision as to validity/infringement and then

---

15   Article 124 (Civil penalties):
      "1. By a judgment ascertaining and declaring the infringement of an industrial property right an injunction from manufacturing, trading and using items infringing the relevant right can be ordered.
      2. When issuing the above injunction, a judge may fix an amount owed for each violation or subsequently ascertained non-compliance, as well as for any delay in obeying the order. [continued overleaf]

subsequently proceed for the damages compensation.

The method for assessing damages is in accordance with the implementation of the EC Directive. Article 125[16] of the IP Code clearly provides multiple criteria:

- lost IP profits;
- reasonable royalties in the alternative; and
- assignment of the infringer's profits.

Most courts are not very generous in terms of awarding damages. Milan, on the other hand, always asserts that "there is no infringement that shall not create damage to be compensated".

The first decision is usually deemed to be final and only rarely will the court of appeal suspend the decision.

10. Costs

The cost assessment is based on the basic principle that the losing party has an obligation to pay the legal fees of the winning party.

Before 2006, judges normally relied on tariffs set out by the National Bar Association. Accordingly the calculation of fees was limited only to the lawyers' fees and did not necessarily reflect actual fees incurred. Moreover, judges always had the facility to reduce the amount officially calculated under the tariffs, at their discretion. The unfortunate consequence of this was that even where a case was successful, the judge would usually grant just a third of the actual fees expended.

---

3. *A judgment ascertaining the infringement of an industrial property right may order the destruction of all infringing items. However, no destruction can be ordered if the destruction of such item is detrimental to the national economy and the person entitled thereto shall only obtain compensation for damages,. In case of infringement of trademarks, the destruction affects the trademark, but can include packaging and if the judicial authority deems it appropriate also products or materials pertaining to the supply of services, insofar as it is necessary to cure the effects of the infringement.*

4. *A judgment ascertaining the infringement of industrial property rights may order that objects produced, imported or sold in violation of the right, including the specific means univocally devoted to manufacture them or to implement the protected method or manufacturing process, be assigned to the holder of the right, without prejudice to the right to compensation of damages.*

5. *A judge may also, if requested by the owner of the items or of the manufacturing means mentioned in paragraph 4 above and taking into account the residual duration of the industrial property title or the particular circumstances, order the seizure at the infringer's expense, until such time as title, items and manufacturing means expire. In such latter case, the industrial property right holder may ask for the assignment in his favour of the items attached at the price to be fixed by the judge for the enforcement after consulting an expert, if the parties have failed to reach an agreement in that respect.*

6. *Items infringing industrial property rights shall not be ordered to be removed or destroyed, and the use thereof shall not be forbidden if they belong to someone making personal or household use thereof.*

7. *A judge who issues a judgment ordering the above measures shall decide on any challenges arising in the course of the enforcement of the measures mentioned in this Article, by a non-appealable order, after summoning the parties and gathering summary information."*

16 Article 125 (Compensation for damages):

"1. *The compensation for damages, owed to the aggrieved party, shall be quantified pursuant to Articles 1223, 1226 and 1227 of the Civil Code, taking into account all the related aspects, as the negative economic consequences, including the lost profits of the IP owner, the profits realized by infringer and, in the appropriate cases, elements not related to the economic ones, as the moral prejudice generated against the IP right owner.*

2. *The judgment ordering compensation for damages may quantify them in an overall amount fixed on the basis of the case evidence and of the presumptions inferred therefrom. In this case the lost profit is in any way determined as an amount not inferior to the royalties that the infringer would have been bound to pay if he had obtained the license from the holder of the right.*

3. *In any case the owner of the infringed IP right may claim the assignment of the infringer's profits, as alternative of the lost profit or to be added to such amount."*

Hopefully, thanks to the entry into force of the Enforcement Directive, article 14,[17] it is now possible (even though there is no specific legal codified rule in Italy) to:

- claim all the fees effectively incurred by the winning party (including the patent attorney's fees); and
- claim the actual amount paid by the winning party.

The judge may reduce these figures in the context of the content of, and reasons for, the decision adopted, but he cannot ignore the values requested.

It is possible to suggest a list of average costs that might be considered reasonable:

- Court fees: €300 (tax for filing an ordinary litigation; *contributo unificato*).
- Litigators' fees:
  - €50,000 to €80,000 for a full proceeding on the merit; and
  - €25,000 to €40,000 for a preliminary proceeding.
- Patent attorney's fees: €7,000 to €10,000 for a party's expertise:
  - €10,000 to €12,000 for a full proceeding on the merit; and
  - €5,000 to €8,000 for a preliminary proceeding.

Of course, these figures should be considered only as averages. Law firms with a strong reputation in the field may require higher fees and less well-known firms may request less.

11. Hot topics

11.1 Harmonisation

One of the hottest topics in Italian patent litigation concerns harmonisation of the position of the various courts.

The fact that there are 12 specialised IP courts means that it is very difficult to achieve consistency of case law. The ideal would be to have fewer than eight courts in order to promote a degree of predictability in decision making.

11.2 Republic of San Marino

In order to give a complete picture of the legal scene, it is worth mentioning that the Italian geographical territory includes the separate State of the Republic of San Marino.

According to an agreement which came into force with effect from March 31 1939, both countries undertook to protect their reciprocal IP rights without the need for parallel registered rights. Since 1999, San Marino has had its own Patent and Trademarks Office.

The internal legislation on IP was finally implemented on May 25 2005 with the adoption of the *"Testo Unico in tema di Proprietà Industriale"*. This regulation now

---

17   Article 14 (Legal costs): *"Member States shall ensure that reasonable and proportionate legal costs and other expenses incurred by the successful party shall, as a general rule, be borne by the unsuccessful party, unless equity does not allow this."*

reflects the content of the TRIPS Agreement – despite the fact that San Marino is not a member of the WTO – and the Enforcement Directive. IP enforcement in San Marino is regulated by the *Testo Unico*, as are the relevant procedural rules.

Some peculiarities of San Marino are:

- recognition of a grace period of six months in order to save the novelty requirement in favour of the patent owner's disclosures before the filing date;
- a common-law system for the litigation issues; and
- the extra-EU position of the country and, consequently, a local "exhaustion" principle only.

The few cases already discussed before the San Marino court of first instance (where the judge is called the *Commissario della Legge*) demonstrate that it has the ability to manage complex juridical issues. Patent litigation in San Marino may, therefore, offer an interesting alternative to typical Italian litigation.

Japan

Masahiro Otsuki
Abe, Ikubo & Katayama

1. Introduction

Patent litigation in Europe varies from one country to another. Japanese patent litigation, however, is quite different from any European jurisdiction, as outlined below.

1.1 Sources of law

Japan is a civil law jurisdiction, which is governed by the constitution as the supreme law, followed by the Civil Code, the Civil Procedure Law, the Patent Law and other statutory laws. In addition to the statutory laws, ministry circulars, local regulations, and Supreme Court decisions are also regarded as sources of law in Japan.

In patent proceedings, although the most relevant statute is the Patent Law,[1] the Civil Procedure Law and the Civil Code are relevant in patent infringement suits, and the Customs Law and ministry circulars are relevant in Customs actions.

1.2 The court system

Japan has a three-tier court system. Patent infringement cases are first heard either by the Tokyo District Court, or the Osaka District Court. These two courts have exclusive jurisdiction over patent infringement cases. Both have divisions specialising in intellectual property cases[2], and a patent infringement case will be assigned to those sections.

A party not satisfied with the district court's decision may appeal to the Intellectual Property High Court (IP High Court).[3] The IP High Court is a special branch of the Tokyo High Court, which has exclusive jurisdiction over appeals from decisions of the Tokyo and Osaka District Courts on patent and utility model litigation, and appeals from decisions of the Japan Patent Office. The IP High Court reviews factual findings and legal determinations made by the district courts. The IP High Court can also conduct factual findings on its own.

Further, a party which is not satisfied with the IP High Court's decision may appeal to the Supreme Court, which is the highest court in Japan. However, unless

---

1. Hereafter, in the body of this chapter, the term "article" refers to articles of the Patent Law in Japan unless otherwise specifically indicated.
2. The Tokyo District Court has four divisions consisting of 17 judges, and the Osaka District Court has two divisions comprising approximately six judges. A panel always consists of three judges.
3. The IP High Court was established on April 1 2005. It has four divisions consisting of nearly 20 judges. Usually, a panel consisting of three judges examines a case. Occasionally, the IP High Court frames a Grand Panel of five judges, when it considers that a case addresses a significant legal issue.

the appeal involves certain issues, the Supreme Court does not review such appeals as of right. Issues to be reviewed by the Supreme Court as of right are strictly limited, but include matters such as violation of the constitution and inconsistency with precedents. In addition, the Supreme Court can review appeals at its discretion when a case includes important legal issues. As a result, only a small percentage of appeals from IP High Court decisions are actually reviewed by the Supreme Court.[4]

The Civil Procedure Law provides[5] that the period for filing a notice of appeal is two weeks from the date on which the lower court decision was served, although a court can allow an additional period at its discretion when a losing party is a foreign company or a person residing abroad.[6]

1.3   Parties involved in patent litigation

(a)   *Legal professionals at courts*

(i)   *Judges*
Law school graduates who pass a bar exam and complete mandatory training can become assistant judges, assistant public prosecutors or attorneys at law (*Bengoshi*). An assistant judge must have ten years' experience in order to become a judge. They are not specialists in specific fields, such as intellectual property. Thus, the percentage of those who also have technical backgrounds is low, even if they engage in the field of patent law.

Judges tend to take note of decisions of higher courts, especially those of the Supreme Court.

(ii)   *The court's technical researchers*
To assist judges in understanding the technological issues involved in patent cases, the district courts and the IP High Court retain court technical researchers (*Saibansho Chosa Kan*).[7] A technical researcher is appointed for each patent case and helps the presiding judges to understand the technical aspect of relevant action.

(iii)   *Outside technical advisors*
Since 2004, the courts have been able to ask an outside technical advisor (*Senmon Iin*) to help judges provide neutral explanations on the advanced and technologically specialised issues involved in patent litigation. Nevertheless, the outside technical advisor's role is quite limited, especially in patent infringement lawsuits.

---

4   There are 15 Supreme Court justices, including the chief justice. Usually, a panel consisting of five justices examines all cases involving patent litigation.
5   Civil Procedure Law, article 285.
6   Civil Procedure Law, article 96(2).
7   The Supreme Court does not have technical researchers, but researching judges (*Saiko-sai Chosa Kan*) who have sufficient experience as lower court judges assist the Supreme Court justices.

(b)   Advocates

(i)   Lawyers (Bengoshi)

Attorneys at law (*Bengoshi*) are allowed to handle any legal matter in any part of Japan. This includes providing legal advice and representing a client in negotiations in all kinds of cases in connection with any form of law. A *Bengoshi* will usually also handle patent litigation. *Bengoshis* can be registered as patent attorneys (*Benrishi*) by filing the necessary paper and fees with the Japan Patent Attorneys Association (JPAA).

(ii)   Patent attorneys (Benrishi)

Patent attorneys (*Benrishi*) are allowed to practise before the Japan Patent Office in all forms of filings in connection with intellectual property rights, including patent rights.[8] A *Benrishi* is also allowed to give an expert opinion on intellectual property rights, handle Customs Office proceedings and alternative dispute resolution (ADR) processes relating to circuit arrangements and unfair competition, as well as the aforementioned intellectual property rights.

Usually, *Benrishis* and *Bengoshis* work together on patent matters, including patent litigation. Generally speaking, the *Bengoshi* will work at persuading the presiding judges in legal matters, and the *Benrishi* will concentrate on explaining technical matters to the judges. Good teamwork between these professionals will be one of the critical factors in achieving success in patent cases.

Benrishis have been allowed to participate in patent litigation as assistant attorneys (*Hosanin*), without the authorisation of judges. In addition, a *Benrishi* who has taken mandatory classes and passed an examination can register as a *Fuki-Benrishi* qualified as an advocate in intellectual property infringement litigation. A *Fuki-Benrishi* can co-represent a client in a patent infringement case, together with a *Bengoshi*.

2.   Claims and defences

2.1   Preparation before bringing a suit

(a)   Patentee's initial reaction

A patentee who finds that a third party is likely to infringe his patent, usually sends a letter to the accused infringer. When it turns out that it is not appropriate to continue such communication/negotiation with the accused infringer, there are various actions (outlined below) that he can take against the accused infringer. The patentee can choose one or more of the following:

- commence a *regular patent infringement* suit with a district court;
- file a petition for *preliminary injunction* (*Kari Shobun*) with a district court;
- file an import *suspension application* (*Yunyu Sashitome*) with the Customs Office; and
- use ADR proceedings such as arbitration and mediation.

---

8   Utility model rights, design registration rights and registered trademark rights are also covered in addition to patent rights.

(i)   *Registered exclusive licensee[9]*
The Patent Law provides two kinds of licence in Japan; one is a registered exclusive licence (*Senyo Jisshi Ken*, article 77) and the other is a regular licence (*Tsujo Jisshi Ken*).[10] The Patent Law provides[11] that a registered exclusive licensee (*Senyo Jisshi Ken-sya*) to a patent can independently sue an infringer for injunction and damages. Since the patentee cannot exploit a patented invention when there is a registered exclusive licensee, it has not been clear whether a patentee can seek an injunction against an infringer. In 2005, however, the Supreme Court held that a patentee is entitled to seek an injunction even if there is a registered exclusive licensee,[12] because the royalty income based on the licensee's sales would decrease unless the infringing act were enjoined.

Hereafter in this chapter, a 'patentee' means a patentee or a registered exclusive licensee, unless specifically indicated otherwise.

(ii)   *Monopolistic regular licensee*
Because a monopolistic regular licensee (*Dokusen Teki Tsujo Jisshi Ken*) does not have a registered exclusive licence, the licensee is not entitled to an injunction under article 100. On the other hand, it is well established that a monopolistic regular licensee is entitled to damages as a result of an infringement.[13]

(iii)   *Non-monopolistic regular licensee*
A non-monopolistic regular licensee (*Hi-Dokusen Teki Tsujo Jisshi Ken*) cannot sue an infringer, either for an injunction or for damages.

(iv)   *Limitation period*
The limitation period for seeking patent infringement damages is three years.[14] On the other hand, a patentee can claim back 10 years based on the unjust enrichment principle, since the relevant limitation period here is 10 years.[15]

(b)   **Patentee's other preparations**

(i)   *Adequate investigation of accused products*
A complaint must contain sufficiently supported substantive arguments showing that the accused infringer's products fall within the scope of the patent claim. Usually, the infringing products are identified by the product name and/or model

---

9   *Senyo Jisshi Ken* is generated by registration of the necessary information on the licence with the JPO (Patent Law, article 98(1)(ii)), and its transfer or other change must also be registered with the JPO. When there is a *Senyo Jisshi Ken*, the patentee cannot practise the patented invention by himself. Furthermore, there are two kinds of regular licence; one is a monopolistic regular licence (*Dokusen Teki Tsujo Jisshi Ken*) and the other is a non-monopolistic regular licence (*Hi-Dokusen Teki Tsujo Jisshi Ken*).
10   The advantage of registering a regular licence is that the registered licence to a patent is effective even if its patentee later transfer the licence to a third party.
11   Patent Law, articles 100 and 102.
12   Judgment rendered on June 17 2005, Supreme Court, *Hei* 16 (*ju*) 997, *Minshu* Vol 59, No 5, p 1074.
13   For example, see judgment rendered on April 27 2004, the Tokyo High Court, *Hei* 14 (*ne*) 4448.
14   Or 20 years from the infringing acts: Civil Code, article 724.
15   Civil Code, article 167(1).

number, and relevant structures of the products in connection with the patent claim should be described. Then, such descriptions of the accused products should be compared with each element of the claim. Therefore a patentee must conduct an investigation into the infringing products, as well as an analysis of the patent at issue, before filing a patent infringement complaint.[16]

Collecting evidence: It should first be noted that there is no discovery proceeding under Japanese law.

If a patent invention relates to articles, the patentee should somehow obtain the suspected products. If the composition and/or effects of the products are at issue, an objective analysis of the products should be carried out before commencing a legal action.

If the patent invention relates to a manufacturing process, the products suspected to be made by the patented process should be obtained for analysis. At the very least, before commencement of a legal action, the patentee should obtain brochures, advertisements, blueprints, and/or photographs of the products and analyse them as to whether they encompass all of the relevant claim elements.

Even if the patent relates to a method including a manufacturing process, the patentee must collect some evidence to establish the patent infringement, such as analysis of the manufactured products, the accused infringer's purchase of raw materials, information from the accused infringer's customers and suppliers, inside informers' information, or technological circumstances.

Document production order: There are two provisions in the Patent Law that assist a patentee in proving patent infringement. The first is article 104 *bis* which provides that an accused infringer must deny the patentee's allegation regarding the infringing products or methods, by clarifying the reasons (*denial with reason*); and the second is article 105, which allows the court to order a party to produce documents (*document production order*). These provisions are, however, not to shift the burden of proof from the plaintiff to the defendant. The accused infringer can also avoid specifying the reason or document production if the court accepts the accused infringer's justification for doing so. In reality, unless the plaintiff submits enough circumstantial evidence to convince the court to make an order, the courts would not allow the plaintiff to use these provisions as convenient tools for searching evidence.[17]

Evidence preservation: Evidence preservation is also available pursuant to Civil Procedure Law, article 234. A patentee can file a petition with a court to order evidence preservation. However, the requirements for these proceedings are extremely hard to meet. The petitioner is required to specify the subject piece of evidence, the fact to be proved by the evidence, and the justification for the evidence preservation. For the justification, the petitioner must demonstrate a prima facie case of infringement as far as is reasonably possible as well as the reason why it is necessary to preserve proof now, rather than later. Once the petitioner obtains

---

16 The plaintiff may be liable for damages suffered by the defendant if the lawsuit is commenced knowingly without factual and legal basis (judgment rendered on January 26 1988, Supreme Court *Sho* 60 (*o*) 122).
17 In Japan, there is no pre-action disclosure system like that of France.

evidence from this procedure, it can be used in other jurisdictions in general.[18]

Other preparations: After collecting evidence, the patentee should expect, and be prepared for, all possible factual and legal issues that are likely to be raised during the course of action. If appropriate, it is a good idea to conduct experiments and prepare an experimental report, and/or consult with an expert on technological issues.

(ii)   Evaluation of patents

Claim constructions: Before commencing the patent infringement suit, the patentee must evaluate the patent at issue. Such evaluation includes studying the patent specification itself and the prosecution history, from filing the patent application through the notice of allowance, at the JPO and any history thereafter, and also a reasonable review of prior art. Evaluating these materials is important for determination of the actual technical scope of the claimed invention. In addition, terms in a claim should be construed in view of the patent specification and the prosecution history of the patent, as well as the common knowledge of a party who is ordinarily skilled in the art. Also, the scope of patent claims may be narrower than is shown on the patent publication, if the patent claims were later corrected through a claim correction proceeding.

Validity of patent: Practitioners should check that the patent at issue is still in force (ie before its expiration date) and that any maintenance fee has been paid.

Further, the prior-art search is important to ensure that the patent at issue is valid for the two following reasons. First, following the Supreme Court *Kilby* decision in 2000 and the legislative affirmation of the decision by article 104ter, patent invalidity is an affirmative defence in a patent infringement suit. Although the courts merely state that the patent is unenforceable and cannot invalidate it at an infringement suit, the patent becomes practically useless since it would be very hard to use it to sue another party based on the same patent. Thus, it is important to identity the weaknesses of a patent in order to estimate the risk of invalidation and be prepared for any likely attacks from the opposing party.

Assigned patent rights: When the patent has been assigned, the assignment is not complete unless it is registered with the JPO.[19] Also, the right to sue based on the patent, before assignment of registration, is not automatically transferred with the patent itself. An explicit assignment of such right is necessary. Moreover, the transfer of such right should be notified to the accused infringer, or his consent to such assignment must be obtained.[20]

Jointly owned patent rights: If more than one person owns the patent, one of the co-owners can seek an injunction acting singly, but cannot do so for the entirety of damages claims. Rather, the co-owner can seek only her quota of the damages.

(iii)   Letter to accused infringer

Letter required for provisional rights: As for the provisional rights associated with

---

18   Japan does not have a pre-action evidence gathering system like that of France.
19   Patent Law, article 98(1)(i).
20   Civil Code, article 467 (1).

a published patent application, pursuant to article 65 (*Hosho-kin Seikyuken*) an actual notice of such patent application is a prerequisite for asserting such rights, unless it can be established that the infringer knew of the patent application.

Warning letters:[21] Although a notice of a patent or even patent marking is not a prerequisite for patent enforcement, sending a warning letter before taking any legal action is a normal course of action in Japan. This course of action can lead directly to licence negotiations.

However, there is a risk that a warning letter might motivate the opponent to take legal action, such as filing a request for invalidation with the Japan Patent Office, or filing a declaratory judgment action. Also, if a patentee sends a carelessly drafted warning letter to a business acquaintance of the accused infringer, the accused infringer may seek damages under the Unfair Competition Prevention Law. In view of such risks, the patentee should be well prepared when sending out a warning letter.

(c)   *Accused infringer's initial reaction*

The accused infringer of a patent infringement suit can take the initiative by:

- filing a request for a patent invalidation trial (*Muko Shinpan*) with the Japan Patent Office (JPO); and/or

- bringing a declaratory judgment suit at a district court, which seeks a declaration of non-infringement and/or unenforceability due to invalidity.[22]

The patent invalidation trial usually takes approximately six to ten months from filing a request to the JPO's decision. Since the losing party can file an appeal with the IP High Court and then the Supreme Court, it takes approximately a further 18 months to two years for the decision to be finalised. The primary objective for requesting the patent invalidation trial is to revoke the patent at issue. The secondary objective might be to prevent the patentee from amending the patent claim through *ex parte* proceedings.

Since appeals from both the JPO and the district court go to the IP High Court, if the district court decision is inconsistent with that of the JPO then the IP High Court can solve any discrepancies by rendering consistent decisions.

In order to bring a declaratory judgment action, an actual dispute must exist between the parties. Moreover, the subject matter for which the declaratory judgment is being sought must be clear. A warning letter which cautions that the recipient of the letter is infringing the patent is usually sufficient proof of a dispute.

(d)   *Accused infringer's other preparations*

On receiving a warning letter from a patentee, the potential defendant should analyse the accused products, conduct a prior-art search, examine the patent

---

21   The letter does not have to be entitled "Warning Letter"; it is often headed "Notice" or "Inquiry".
22   In turn, the patentee can file a countersuit (*Hanso*) seeking damages and/or an injunction in response to the declaratory judgment action. The counterclaim will be examined together with the declaratory judgment action; however, the court usually recommends that the plaintiff of the declaratory judgment action withdraw the complaint because the declaratory judgment action is considered to achieve its objectives by the patentee's filing of the counterclaim.

(including its prosecution history) and obtain a certified copy of its registration. Such actions will assist in devising a meaningful response to the warning letter.

2.2 Procedure and timescale of proceedings

(a) Overview of proceedings

The average duration of infringement proceedings has been significantly shortened in the past decade, although there is no accelerated proceeding. Currently, the court deliberation period for intellectual-property-related cases is less than a year at the Tokyo District Court. Most cases end within 18 months.

After a plaintiff files a complaint with a district court, the court serves the complaint on the defendant. Simultaneously, the court sets the date for the first oral hearing and the due date for the defendant's answer, and notifies the defendant of these dates on the service of complaint. The first oral hearing date is normally set approximately one month to one and a half months after service, and the due date for submission of an answer is set about one week prior to the first oral hearing date.

The first oral hearing will be held in a courtroom and will be open to the public. A panel consists of three judges. The parties formally put forward supporting documents submitted as evidence, together with the complaint or the answer.

There is no discovery or concentrated trial under the Japanese law. Rather, the court will examine the case through a series of hearings held periodically.

In patent infringement lawsuits at the Tokyo District Court, the second and subsequent hearings are held as preparatory proceedings (*Benron Junbi Tetsuzuki*). Preparatory proceedings are closed-door hearings held in a court conference room rather than a courtroom.

In preparatory proceedings, points of dispute are narrowed down and substantive arguments are made by exchange of briefs and evidence. Typically, either or both parties will be instructed to submit counterarguments to the other party's arguments by the next hearing.

It should be noted that, the court virtually divides infringement proceedings into two stages, although it is still formally a single action. The first stage is for infringement and invalidity issues and the second relates to damages issues. The proceedings move to the second stage only when the court finds an infringement and non-existence of grounds for invalidation of the patent at the first stage. Therefore, the parties first focus on arguments relating to infringement and invalidity issues. When moving on to the second stage, the court requests the defendant voluntarily to submit materials showing sales figures such as bills of sales, profit and loss statements, monthly sales books and purchase ledgers. If the defendant does not comply with such requests, the court may issue a document production order (see Section 2.2(b)(v) below), at the request of the plaintiff.

When the court reaches a tentative conclusion, especially in cases where it considers that the patent is likely to be infringed and also that the patent is not likely to be invalidated by the Trial Boards of the JPO, it will often seek a judicial settlement.

If the parties agree to start settlement discussions before the court, the court will commence judicial settlement proceedings (*Wakai Tetsuzuki*). At the beginning of the judicial settlement proceedings, the court often discloses an unofficial tentative conclusion to the parties and informs them that it will be rendered as the actual decision if the parties do not reach a settlement agreement. This tentative conclusion disclosed by the court is usually the same as the final conclusion set out in a judgment. Therefore, the parties will be clear as to the possible outcome, when determining whether they should settle the case or obtain a judgment.

The diagrams below show the standard schedule of patent infringement lawsuits at the Tokyo District Court[23] and the Osaka District Court.[24] The plaintiff's choice of court for filing a lawsuit basically depends on the defendant's corporate address and the place of infringing activity.

Flowchart of typical infringement litigation at the Tokyo District Court

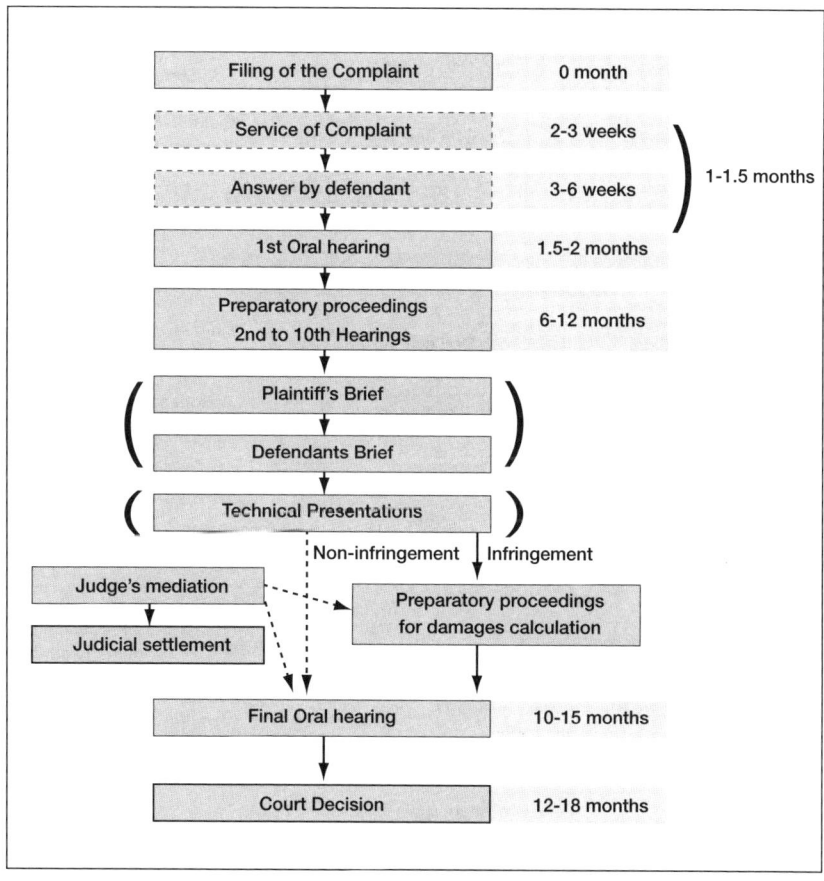

---

23   It is not issued by the Tokyo District Court, but is meant to be a standard Schedule of the Court.
24   The Standard Schedule of the Osaka District Court is based on the "Examination Model of Patent and Utility Model Infringement Cases" issued by the Intellectual Property Divisions of the Osaka District Court (final revision, April 2004).

Standard schedule for patent infringement litigation at the Osaka District Court

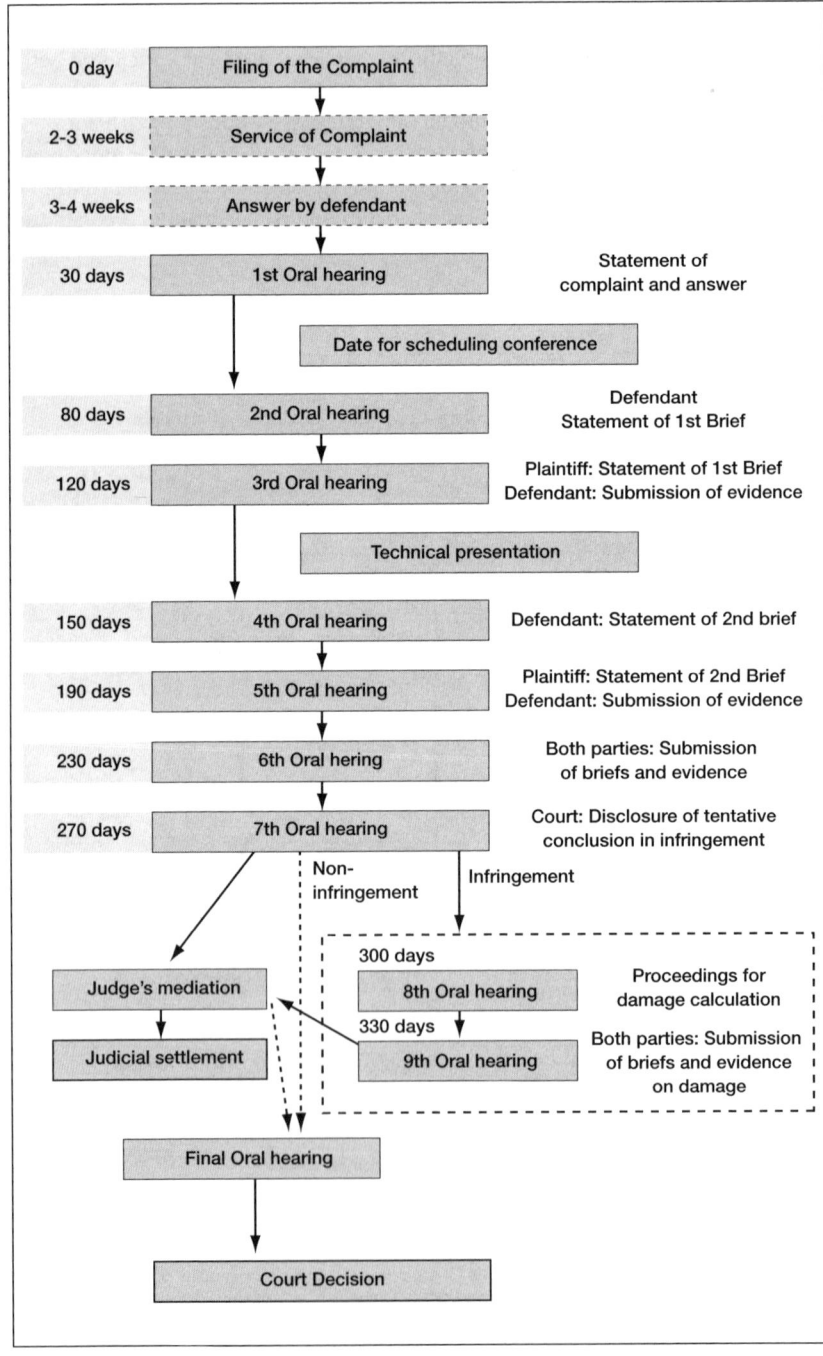

(b)   Essential features of the proceedings

Some of the essential points of the proceedings are set out below.

(i)   Brief

First, a patentee must submit a complaint. A complaint must contain, both parties' names and addresses and those of the plaintiff's attorneys, along with the "gist of demand" and "grounds for demand". Service of a complaint is carried out *ex officio* by the court. If the defendant is a foreign company not having any branch office or place of business in Japan and is located in a foreign country, the court will commission government offices of the foreign country, or the Japanese ambassador in the country, to serve the complaint to the defendant. In such cases, the plaintiff must prepare a translation of the complaint in the language of the relevant foreign country.

In response, a defendant submits an answer. In an answer, the defendant is supposed to admit, deny, or state that he is unaware of, each and every fact asserted by plaintiff in the complaint. In addition, the answer provides counterarguments to the plaintiff's arguments made in the complaint. Unlike a complaint, an answer and other briefs may be served without the involvement of the court (ie sent directly between parties).

After this, the parties exchange briefs with evidence throughout the litigation. Typically, if one party submits its brief with evidence in a hearing, the other party submits its brief in turn rebutting the opponent's arguments in the next hearing. Parties repeat such submission of briefs with evidence. In practice, the court instructs which party should submit a brief of what arguments, and by when.

(ii)   Evidence

Generally, the court prefers documentary evidence,[25] and witness examination is rarely allowed in patent infringement litigation.[26,27]

Since there is no discovery procedure in Japan, collection of evidence before filing of a complaint is very important. Generally speaking, it is more difficult to establish infringement process claims (especially claims describing a process of manufacturing), than product claims.

(iii)   Preparatory proceedings (particularly in the Tokyo District Court)

Usually, among the three judges comprising a panel, one judge assigned by the panel always attends and conducts preparatory proceedings. Furthermore, the presiding judge often attends preparatory proceedings. Courts' technical researchers (*Saibansho Chosa-kan*), outside technical advisers (*Senmon Iin*) (if any), and court clerks also attend hearings of preparatory proceedings. Directors, employees and other relevant people connected with the parties to the litigation may sit in on hearings, if permission is granted by the court (and such permission is usually granted).[28]

---

25   Actual experiments are seldom held before courts.
26   On the other hand, in litigation in which an employee seeks compensation for his invention, it is common to hold examinations of witnesses and of the parties themselves. This is also the case where a prior-use defence is raised by a defendant.
27   Expert examination is also rare. Experts are usually instructed by the parties.
28   Civil Procedure Law, article 169(2).

In preparatory proceedings, parties formally state their briefs and submit supporting documents and materials as evidence. In these proceedings, judges typically address questions to parties in order to clarify the arguments set out in their briefs and establish the kinds of counterargument the other party will make in the next hearing. Such questions and comments will often hint at the judges' opinions on particular issues, and experienced lawyers will be sensitive to the implications and build further strategies. At the end of a hearing, the court sets the next hearing date and the due date for submission of briefs.

At the first oral hearing after conclusion of the preparatory proceedings, the parties formally state the result of the preparatory proceedings.[29] At this oral hearing, the court usually closes the proceedings and announces the date for rendering its decision.

(iv)   *Technical presentations*

The court sometimes requests the parties to make technical presentations. Such presentations are often made at the beginning or end of the infringement and invalidity stage. The objective of technical presentations is for the court to clarify the arguments made by the parties and confirm that the court has correctly understood the issues and arguments of the parties.

(v)   *Document production orders and protection of confidential information*

Although there is no discovery in Japan, the Patent Law[30] provides that an infringement court can order a party to produce documents at the other party's request for the purpose of establishing infringement activities or calculation of damages, unless there is a good reason for the holder of the subject documents to refuse production of the documents.[31] In order to determine whether there is such a good reason for refusal, the court can examine the documents in camera.[32] In reality, however, the courts rarely issue document production orders.

"Protective orders to protect confidential information" were introduced by the amendment of the Patent Law in 2004.[33] Upon a party's request, a patent infringement court[34] may issue a protective order to briefs and evidence containing confidential information of the party, such as information relating to the defendants' process/products and other trade secrets. When a protective order is issued, the party in receipt of the confidential information may neither use it for purposes other than the subject litigation, nor disclose such confidential information to third parties.

Although this new system of protective orders has been introduced, it has not yet been widely used. There have been only a few cases where courts actually issued

---

29   Civil Procedure Law, article 173.
30   Patent Law, article 105.
31   Patent Law, article 105(1).
32   Patent Law, article 105(2).
33   Patent Law, article 105 *quater.*
34   Protective orders are also available in unfair competition litigation relating to trade secrets, such as the defendants' know-how and customer lists.

protective orders.[35] Instead, parties tend to choose voluntary non-disclosure agreements.

(vi)   Restriction of access to judicial records

Access to judicial records by third parties may be restricted by the court at the request of a party, in cases where such judicial records contain trade secrets of the party under the Civil Procedure Law.[36]

(vii)   Conclusion of examination and decision

When the panel considers that both parties have submitted all arguments and evidence and the case is not appropriate for judicial settlement, it concludes a series of preparatory hearings and holds a court hearing that is open to public. At the hearing, the panel confirms all the proceedings taken during the preparatory hearings and declares the conclusion of the examination of the case, unless any issues are raised to be discussed further. At this hearing, the panel also states the date by which it will render a decision.

A district court decision consists of:

- the main element; and
- facts and grounds.

The main element is almost identical with the gist of the demand, either in the complaint or the answer.

(viii) Appeal to upper courts

A party not satisfied with the district court's decision may appeal to the IP High Court and, if still not satisfied with the IP High Court's decision, also appeal to the Supreme Court. However, only a small number of cases are actually reviewed by this forum.

(ix)   Legal fees and costs

Court fees: A complaint must describe the value of subject matter and be accompanied by revenue stamps (*Inshi*) for the relevant court fees. Court fees are determined in accordance with the value of the relevant subject matter of the claim.

Examples of court fees are provided below:

Subject matter value (million Yen)	Court fees (thousand Yen)
10	50
100	320
1,000	3,020
10,000	16,020

---

35   Judgment rendered on September 15 2006, Tokyo District Court, *Hei* 18 (*mo*) 9933, *Hanji* Vol 1958, p 131; a case where a protective order was issued with respect to materials attached to an application for import licence for a generic drug.

36   Civil Procedure Law, article 92.

Attorneys' fees: Some attorneys[37] have recently charged by time actually spent working on a specific matter in patent litigation. Hourly rates vary based on each attorney's experiences and ability, approximately in a range of Y10,000 to Y100,000. **Fees for consultants, experts or other witnesses:** Any arrangement is possible if agreed between a party and a consultant (ie lump sum amount or periodical payment based on time actually used).

Costs orders: Generally, a court decision includes an order that the losing party pay the legal costs.[38] The 'legal costs' means court fees rather than the entire costs of litigation. Although a court may order a losing party to pay some portion of the winning party's attorney's fees and costs as a proportion of damages incurred to the wining party, the relevant amount of attorney fees will usually be at most 10% of the damages award.

2.3 Remedy in regular patent litigation

A patentee whose patent is infringed may seek remedies, mainly injunctions and monetary compensation. Injunctions are available to eliminate present and future infringements.[39] For past infringements, a patentee can be awarded damages,[40] or compensated based on the doctrine of unjust enrichment.[41] These remedies can be sought separately or together, and a patentee may consider what remedy or remedies will most effectively cure the wrongful acts that the infringer has committed.

(a) Injunctions

Article 100(1)[42] provides the right of a patentee to demand an injunction.

(i) Requirements for injunction

A patentee may seek a permanent injunction against present or future infringements, even if the infringer is neither wilful nor negligent.[43]

To obtain an injunction, a patentee must specify the nature of the activities of the infringer to be stopped or refrained from. Also, if the subject of the injunction is a future infringement, the patentee must show that future infringement is both likely and imminent. This likelihood must be objectively decided, based on an examination of the infringer's intent and the stage of preparation of the infringement.

(ii) Acts to be banned by injunction

If the patent is on an invention of a product, acts such as manufacture and sale of infringing products can be banned; and if the patent is on an invention of a process of manufacturing a product, then both the use of such a patented process and acts such as use and sale of the products made by the patented process can be banned.

---

37   Usually those working with large law firms which numerous foreign clients.
38   Civil Procedure Law, article 61.
39   Patent Law, article 100.
40   Civil Code, article 709 and Patent Law, Art 102.
41   Civil Code, articles 703 and 704.
42   Article 100(2) provides a patentee the other measures to reinforce the effect of injunction.
43   Patent Law, Art 100(1).

On the other hand, if the patent is on an invention of a process, a patentee can demand an injunction only for using such a patented process. Therefore, once a patented invention is recognised as an invention for a simple process, it would be difficult for a patentee to seek an injunction relating to the manufacture or sale of products relevant to such an invention.

(b) Damages

(i) General

Where an act is found to be infringement of a patent, such act constitutes a tort, which is prescribed in article 709 of Civil Code, and the patentee may demand compensation for damages.

(ii) Compensation for infringement before patent registration

As a patent will come into force upon registration, a patentee may seek damages for infringing acts *after* the patent registration, as compensation based on tort. However, after a patent is granted and registered, a patentee may request royalty-based compensation from the third parties who practise the inventions of the patent application during the period from the date when the application was laid open[44] to the date when the patent is registered.[45] It should be noted that the patentee must give a written warning to the aforementioned parties before making such a demand, unless it can be established that the infringer knew of the patent application.

(iii) Presumptions on demand for damages

Under the law of tort, a patentee can obtain compensation for damages where he can show:

- that the infringer is wilful or negligent;
- that the infringing act has caused the patentee to suffer a loss; and
- the amount of financial loss suffered as a result of the infringement.[46]

To cope with the difficulty of proving such facts, however, the Patent Law provides presumptions for two aspects of the above requirements. First, "an infringer is presumed negligent in the commission of [his] act of infringement".[47] Secondly, the Patent Law prescribes the presumption of the amount of damages.

Presumption of the amount of damages: Proving the amount of damages under the law of tort is generally difficult. Thus, a few important formulas are provided to lessen the challenge faced by patentees' in this area (see article 102):

- **Infringer's sales:** A patentee may seek damages based on an infringer's sales of infringing products with his own profit rate, based on article 102(1). In addition, a patentee will generally wish to assert that he has lost the

---

44   Under the Patent Law, article 64, the content of the patent application is automatically laid open to the public for 18 months after the filing date of the application (or 18 months after the filing date of the first application in the originating country when one or more priorities have been claimed).
45   Patent Law, article 65(1).
46   Civil Code, article 709.
47   Patent Law, article 103.

opportunity of selling his products by the quantity of the products an infringer has sold and that the amount of damages should be the profit for the quantity of the products he could have sold but for such infringement. Under article 102(1), a patentee may claim as damages the amount of the profit corresponding to the infringer's sales, disclosing and using the patentee's own profit rate, if the patentee is able to disclose those factors. As for proving an infringer's sales, under article 105 it is possible to claim an order to produce documents, so as to make the infringer disclose data relating to its sales and profits. The formula under article 102(1) is an effective tool for the plaintiff if he is willing to disclose the profit rates of her own products;

- **Infringer's profit:** Based on article 102(2), a patentee may also seek damages based on the presumption that an infringer's profit is the patentee's damages; and

- **Licence royalty:** Under article 102(3), a patentee is allowed to demand compensation for damages based on the appropriate royalty rate. If the patentee can prove that the damage suffered was more than he would have received as a normal royalty, then he can demand to be compensated accordingly.[48] Where an infringer is not wilful, or does not infringe a patent with gross negligence, the court can reduce damages at its discretion.[49] However, even in such cases, the damages cannot be less than what would be considered as a reasonable royalty. In order to prevent infringers adopting a strategy to the effect that "it is better to pay a small amount of damages in the suit, rather than to avoid being sued", courts' findings on royalty rates are becoming increasingly favourable to patentees.

Range of damages to be compensated: There are a number of factors that have to be taken into consideration, as follows:

- **Consideration on contribution of patented invention:** Where a patented invention pertains to a component of a final product, the patentee cannot necessarily collect all the infringer's profits. In many cases, patentees will be limited to collecting the infringers' profits to the extent that the patentees' inventions have contributed to the infringers' products;

- **Costs, expenses and lost profits:** A patentee may seek as damages not only lost profits, but also costs and expenses actually suffered, such as costs for examinations on infringing products;

- **No punitive damages for wilful infringement:** There is no rule in Japan that allows a patentee to demand more than the actual damages suffered, even if an infringer has behaved wilfully. On the other hand, the amount of damages should not be reduced even if the plaintiff has failed to fix patent markings on the patented products; and

- **Infringement by multiple infringers:** Where a patent is infringed by multiple infringers and these infringers have some relationship such as

---

48 Patent Law, article 102(4) (corresponds to former article 102(3)).
49 Patent Law, article 102(4).

'manufacturer and seller' or 'wholesaler and retailer', the acts of infringement by them are considered joint torts (*kyodo fuho koi*). As such, the infringers will be jointly and severally liable.[50] The same rule applies to cases where infringers take partial charge of the act of infringement, such as manufacture.

Expert damages calculations: At the request of a party, the courts may appoint an expert to calculate damages, and both parties must provide the expert with the necessary information for the expert opinion to be given.[51]

Award of reasonable damages: If the patentee shows the presence of damages but cannot prove the amount of the damages because of the nature of the relevant facts in the case, the court may determine the reasonable amount of damages at its discretion, based on the entire record of oral proceedings and the examination of evidence.[52] This provision is expected to provide a patentee with appropriate compensation, when it is difficult to prove the amount of damages.

2.4   Interim relief (preliminary injunctions)

Preliminary injunction proceedings are independent proceedings which are separate from a regular infringement suit. To obtain a preliminary injunction order, the patentee must establish:

- an infringement; and
- the necessity for a preliminary injunction[53] (ie the possibility of suffering irreparable harm).

Patent invalidity is an affirmative defence.

A preliminary injunction proceeding is different from a regular lawsuit in several ways. First, by definition, a preliminary injunction does not result in an entitlement to recovery of damages. It is purely an injunction. The case is examined by a single judge and the proceedings are not open to the public. The proceedings take less time than a regular lawsuit. It usually takes approximately five to ten months for a district court to issue a preliminary injunction order. The proceedings are conducted *inter partes* and are similar to those for regular infringement suits. The court usually holds a series of court conferences until it considers that the necessary arguments and evidence have been presented. The proceedings may take longer, however, if the case involves complicated issues. The patentee is usually required to deposit a sum of money as security for obtaining the order.[54] Once a district court issues the preliminary injunction order, the preliminary injunction can take place; although the accused infringer can file an objection, it cannot stay the injunction.

---

50   Civil Code, article 719(1).
51   Patent Law, article 105[bis.]
52   Patent Law, article 105[ter.]
53   It would be difficult to show the necessity for a preliminary injunction, if either neither the patentee nor its licensee practice the patent invention in Japan.
54   The amount to be deposited is determined by the court in consideration of the damages that the accused infringer would incur and the likelihood that the order later turns out to be wrong. The security deposit will be returned when the infringement court upholds the injunction and the decision becomes final, or when the opponent gives consent.

The court fee for a preliminary injunction is inexpensive (ie uniformly ¥2,000 per action). As stated above, however, the court will usually order the patentee to deposit a certain amount as a security before issuing a preliminary injunction order.

2.5 Settlement

(a) Judicial settlement

Approximately 50% of intellectual property suits end with judicial settlements. Usually, when a panel has reviewed the entire case on infringement and patent validity issues (ie approximately six to ten months after the commencement of litigation), it takes the initiative on settlement negotiation by disclosing, or sometimes vaguely suggesting, to each party its unofficial tentative determination on the infringement and patent validity issues in question.

The advantages of judicial settlement differ depending on whether or not the court's preliminary determination is favourable. If the court's preliminary unofficial determination is favourable, the advantages of judicial settlement lie in being able to bring an early end to the proceedings and thereby avoid the monetary burden of an appeal and non-monetary risks such as a patent invalidity decision by the JPO and/or a court if the case does not settle. It is also advantageous that the settlement can include more conditions than the decision, if appropriate. Parties can ask the court not to make the details of a settlement agreement open to the public.

If the court's preliminary unofficial determination is unfavourable, the advantages of a judicial settlement are, in addition to those mentioned above, to be able to avoid an unfavourable decision in full proceedings. This in turn might mean that a smaller payment might result than in an actual decision in full proceedings.

(b) Non-judicial settlements

A typical example of a non-judicial settlement is one achieved through alternative dispute resolution (ADR) proceedings, such as arbitration and mediation.

3. Substantive issues of patent litigation

3.1 Existence and survival of patent rights

Any discussion on patent infringement must begin with an enquiry as to whether patent rights exist in the first place. Article 66(1) provides that patent rights in Japan are rights created by the Patent Law and do not exist unless and until the registration of the patent grant has been made.

Once patent rights come into existence on registration, they continue to be valid and effective until:

- expiry of the patent term (ie 20 years from filing of the patent application);[55]
- they are invalidated by a decision of the JPO[56] in invalidation trial proceedings;

---

55 It can be extended for up to five years under limited conditions.
56 Although Japanese courts today may find that the patent (claim) in the infringement suit has a ground for invalidation and dismiss the suit for this reason, it is neither a declaration nor determination of invalidity of the patent, which is still within the exclusive authority of the JPO.

- they are extinguished because a patentee fails to pay the maintenance fee.[57]

3.2   What is patent infringement?

(a)   Elements of patent infringement

Based on the implications of article 68, a patent infringement can be defined as 'an unauthorised practice of the patented invention as a business'. The meaning of 'practice' is defined in the Patent Law depending on three types of invention:

> *"Article 2(3) 'Practice' of an invention in this Law means the following acts:*
>
> *(i) in the case of an invention of a product (including the program, etc, hereinafter the same), making, using, assigning, etc (which means assigning and leasing, and if the product is program, etc, includes providing via telecommunications circuit – hereinafter the same), exporting or importing or offering for assignment, etc (including displaying for the purpose of assignment, etc – hereinafter the same) of the product;*
>
> *(ii) in the case of an invention of a process, using the process;*
>
> *(iii) in the case of an invention of a process of manufacturing a product, using, assigning, etc, exporting or importing or offering for assignment, etc of the product manufactured by the process, in addition to the acts mentioned in the preceding paragraph."*

Interpretative issues on some terminologies used in this article are explained next.

'Offer for assignment, etc': This term includes display, distribution of catalogues and pamphlets, and solicitation to this end.

'Manufacturing process': As regards inventions directed to a process for manufacturing a product, 'practice' involves not only acts of using the process but also practices for the product manufactured by the process, such as importing and selling it. Thus, even if a product is manufactured by the patented process outside Japan, importation of the product into Japan constitutes a 'practice' of the patented process.

In connection with 'using' a manufacturing process, the patentee usually faces a difficulty in establishing that the defendant's process is identical to the patented process, because there is no system of discovery in Japan. For reducing this difficulty, however, article 104 provides the presumption in relation to the defendant's manufacturing process. Under this provision, a patentee of the manufacturing process can shift the burden of proof for establishing infringement onto the defendant, by showing that:

- an object of the patented process was novel at the filing of the patent application; and
- such object and the defendant's product are the same.

If the burden is so shifted, the defendant will be held liable for infringement

---

57   There exist other factors to extinguish patent rights, such as relinquishing a patent.

unless he discloses his process to the court and establishes that the process is different from the patented process.

'Making' and 'reconstruction/repair': As regards 'making', it is generally said that 'reconstruction' constitutes 'making' and 'repair' does not. However, the line between impermissible reconstruction and permissible repair is difficult to draw. This issue often arises in relation to 'exhaustion' of a patent right.

'As a business': It is generally agreed that this term represents conduct which is part of economic activities.

(b) *Direct infringement*

Direct infringement is found where an infringer conducts an act defined as 'infringement' within the above meaning.

To link an accused infringer's act to the patented product or process, it is necessary that the elements of an accused infringing product or process meet all the constituent elements of the claim of the patent. Where there is dispute between the parties as to the meaning of a constituent element, claim construction should be examined. On the other hand, an unauthorised commercial performance of any act is infringement, even if it is committed without knowledge of the patent.

Manufacture, use or sale outside Japan is not infringement in principle. However, the importation and sale in Japan of products manufactured abroad by a patented process is infringement of a patent of such a process.

(c) *Indirect infringement*

Where certain acts do not directly infringe a claim but are preliminary or contributory to infringement, they are likely to induce an infringement if allowed to stand, and it would be difficult to prevent the infringement once they are committed. Thus, such acts are deemed to be acts of infringement,[58] which can be subject to an injunction and/or a demand for compensation for damages. An act falling within this category is called "an act deemed to be an infringement" under the Patent Law and is also referred to as an indirect infringement. For a certain act to be an indirect infringement, it is not necessary for someone to commit an act, as in the case of direct infringement. The details of indirect infringement are stipulated in article 101.

3.3 Offence of patentee infringement

(a) *Overview*

The main issue in infringement litigation is whether the accused product or process is identical to the patented product or process. First, the claim allegedly infringed is construed for ascertaining "the technical scope of the patented invention" set out in the claim. Second, the construed claim is applied to the accused product or process. If the accused product or process is covered by the claim, such product or process is identical to the patented product or process. This is called 'literal infringement'. Even if, however, the accused product or process is not literally covered by the claim, a

---

58   Patent Law, article 101.

patentee may sometimes win the lawsuit because such product or process is found to be substantially identical to the patented product or process under the doctrine of equivalents.

(b) Literal infringement – technical scope of the patented invention

(i) "Technical scope" – article 70(1)

The key provision for the literal infringement is article 70(1). This states, "The technical scope of the patented invention shall be determined based on the description of the patent claim(s) attached to the application."

In general, the technical scope of the patented invention is the same as the scope of the claim.

(ii) General principles

Article 70(1) leads to the following two general principles:

- What is described in the claim will be considered as an element or limitation that constitutes the patented invention; and
- What is not described in the claim will not be considered as an element or limitation that constitutes the patented invention.

(c) Finding the technical scope – claim construction

A process of claim construction will be necessary in order to ascertain the technical scope of the patented invention. The description of a claim can be broken down into elements (or limitations) of the patented invention. Thus, claim construction is substantially the same as the construction of elements of the patented invention.

Under article 70(2), descriptions of the specification and drawings are always taken into consideration for construing the elements (ie terms) of the claim. This must be done, regardless of whether or not the claim terms are ambiguous.[59]

Terms of the claim are generally given their ordinary and customary meaning. The ordinary and customary meaning of a claim term is the meaning it would have to a person of ordinary skill in the art. It is important, however, that terms of the claim do not stand alone and are construed in the context of the specification as a whole.

(i) Specification

The specification is the most important guide to the meaning of a disputed claim term. The underlying rationale is that the description of the specification must support the claimed invention and enable any person who is ordinarily skilled in the art to carry out the claimed invention.

However, if there is a discrepancy in scope between the description of the patent claim and the description of the specification, the former will prevail.

Also, in construing claim terms, courts will usually not read limitations from the

---

59 It should be noted that this theory is applicable to the claim construction for infringement analysis. On the other hand, most scholars and practitioners take the view that for ascertaining invalidity analysis as well as patentability, a detailed explanation of the invention in the specification can be taken into consideration only in limited circumstances.

specification into the claims. Thus, in general, even if a patent describes only a single embodiment, the claims of the patent will not be construed as being limited to that embodiment.

The "detailed explanation of the invention" in the specification generally consists of the following parts:

- "technical field of the invention";
- "prior art and technical problems to be solved";
- "means for solving the problems";
- "embodiments"; and
- "meritorious effects".

(ii)   *Prosecution history*

Japanese courts almost always construe claims in light of the prosecution history of the patent if this is presented by the parties. It can inform the meaning of the claim language by demonstrating how the inventor understood the invention and whether the inventor limited the invention in the course of prosecution, making the claims scope narrower than it would otherwise be. If a claim term is construed more narrowly than it would otherwise be in light of its prosecution history, it is often called an application of the 'prosecution history estoppel' or 'file wrapper estoppel'. In Japan, it seems to be agreed that the prosecution history estoppel may apply to literal infringements as well as infringements under the doctrine of equivalents.

(iii)   *Prior art*

Prior art may be considered for construing claim terms. In current practice, consideration is given to the purpose of recognising the technical level of the relevant art at the time of the filing of the patent application, and thus for finding what a person of ordinary skill in the art would understand the disputed claim terms to mean at such time.

(iv)   *Dictionaries and technical papers*

Dictionaries and technical papers may be used for claims construction. They are taken into consideration for finding general and customary meanings of the claim terms, as well as the common knowledge of a person of ordinary skill in the art. However, when there is a discrepancy between the ordinary and customary meaning of a claim term understood from dictionaries and that understood from the context of the specification, the latter will prevail.

(v)   *Expert opinions*

Generally, judges in infringement courts specialised in patent laws are not interested in hearing expert opinions regarding claims construction.

(d)   **Infringement under the doctrine of equivalents**

(i)   *Supreme Court's* Ball Spline *decision*

For a number of years, there was some debate as to the issue of whether Japanese courts adopted infringement under the doctrine of equivalents and, if so, under what requirements. The Supreme Court ruled positively on the issue in the so-called *Ball Spline* case in 1998.[60] The court set out five requirements for finding infringement under the doctrine of equivalents where literal infringement cannot be found:

- **Non-essential part:** Parts of the claim which are different from the structure of the accused product or process ('accused embodiment') are not essential parts of the patented invention;
- **Interchangeability:** The accused embodiment achieves the same purpose and produces the same meritorious effect as those of the patented invention, even with the existence of the different part;
- **Obviousness of interchangeability:** The interchangeability in the second requirement was obvious to a person ordinarily skilled in the art at the time of manufacturing the accused embodiment;
- **Accused embodiment not falling within public domain:** The accused embodiment is not the same as, or not obvious from, the prior art at the time of filing of the patent application; and
- **No special circumstances:** There are no special circumstances to deny the infringement, such as the intended exclusion of the accused embodiment from the scope of the claim in the prosecution history, etc.

Following the *Ball Spline* decision, the requirements for the doctrine of equivalents are now clear; however, there have not been significant numbers of patentees who have succeeded in their claims as a result of this theory. There also remain many issues for further discussion regarding the five requirements for equivalency.

3.4 Accused infringer's defences

(a) Non-infringement ("non-satisfaction of element") defence

The non-infringement defence is the main defence in Japanese patent litigation. It should be noted, however, that the non-infringement defence is, in a strict sense, a rule of civil procedure, not an 'affirmative defence', because the burden of proof of infringement is always on the patentee.

To make a non-infringement argument, the defendant very often relies on the disclosure of the specification, prosecution history, and/or prior art, arguing that any one or more of the elements of the patented invention should be construed in such a way that the accused product or process does not satisfy such element.

If infringement under the doctrine of equivalents is at issue, the defendant positively asserts that the accused product or process does not satisfy any one of five requirements set out by the *Ball Spline* decision. Notwithstanding the aforementioned comment on the burden of proof, the fourth requirement and the fifth requirement of the doctrine of equivalents (see bullet points above) are exceptions; an accused infringer must prove non-application of these requirements.

---

60   Judgment rendered on February 24 1998, Supreme Court, *Hei* 6 (*o*) 1083, *Minshu* Vol 52, No 1, p 113.

(b)   *Invalidity defence*

Currently, the invalidity defence is the most common affirmative defence argued by accused infringers in infringement litigation. Historically, this defence was not admitted in Japanese patent litigation. The Supreme Court, however, decided in 2000 in *Fujitsu v Texas Instruments* (the *Kilby* decision),[61] that a patent cannot be enforced under the theory of abuse of rights if it is clear that the patent has a ground for invalidation, even though the patent has not been invalidated by the JPO.[62] The *Kilby* decision held:

> *"even before the trial decision to invalidate the patent has been made final, the court that examines the patent infringement suit may decide whether the patent has a clear ground for invalidation. If it is clear, as a result of the examination, that the patent has a ground for invalidation, the claims for injunction and damages, etc under such patent rights are not allowed, unless there are special circumstances, since it constitutes the abuse of rights."*

As a result of the *Kilby* decision, an argument based on patent invalidity, or (more accurately) based on the "clear ground for invalidation", has become a major defence in patent infringement litigation.

In this trend, a new provision, article 104*ter* (1) was introduced by the amendment of the Patent Law in 2004, which states that a patent cannot be enforced in an infringement action if it is found that the patent is to be invalidated in an invalidation trial. The grounds for invalidity are not limited, so long as they are available in the invalidation trial, and they include lack of novelty under article 29(1), lack of inventive step under article 29(2), and failure to meet the disclosure requirements regarding the claim or the specification under article 36.[63]

(c)   *Relationship between the invalidity defence and correction of claims*

If the invalidity defence is raised by a defendant, the plaintiff will often wish to amend the attacked claim to avoid invalidation of the patent. This is 'the trial for correction' or 'the demand for correction in the invalidation trial'.

A correction of the claim demanded in the invalidation trial will become effective when the JPO decision on the invalidation trial becomes final. Due to the possibility that the trial decision of the JPO can be appealed to the IP High Court in the same way as the revocation suit, it will generally take over a year for the correction to become final. On the other hand, the trial for correction will generally take three to four months, and the correction becomes final when the trial decision is rendered.[64]

(d)   *Other defences*

A defendant can argue other issues in patent litigation, such as prior use under article 79, experimental practice under article 69, or exhaustion of patent rights.[65]

---

61   Judgment rendered on April 11 2000, Supreme Court, *Hei* 10 (o) 364, *Minshu* Vol 54, No 4, p 1368.
62   Thus, even if a defendant files a request for a patent invalidation trial with the JPO during pending the patent litigation, the infringement court rarely stays its proceeding.
63   Detailed treatment of novelty or inventiveness in Japan is beyond the scope of this chapter.
64   The trial, however, can be filed in limited conditions.
65   The Supreme Court has rendered important rules on all issues; judgment rendered on October 3 1986, Supreme Court, *Sho* 61 (o) 454, *Minshu* Vol 40, No 6, p 1068; judgment rendered on April 16 1999, Supreme Court, *Hei* 10 (*ju*) 153, *Minshu* Vol 53, No 4, p 627; judgment rendered on November 8, Supreme Court, *Hei* 18 (*ju*) 826, *Hanrei Times* No 1258, p 62.

4.   Useful resources for non-Japanese speakers

There are many sources of information on Japanese patent litigation available in Japanese, such as statutory laws, case laws, websites of public or private organisations, academic books or articles and lectures by professionals. Unfortunately, however, there are few reliable sources available in foreign languages. Examples of such sources are listed below.

4.1   Statutory laws

Most of the relevant statutory laws have been translated into English, and are available on Japanese government websites (http://www.cas.go.jp/jp/seisaku/hourei/data2.html). It should be noted, however, that these are unofficial translations. Only the original Japanese texts of the laws have legal effect, and the translations are to be used solely as reference material to aid in understanding Japanese laws. Moreover, the Patent Law is often revised, and translations may not reflect the most up-to-date version. Despite the drawbacks, it is still an important tool for understanding Japanese patent litigation.[66,67]

4.2   Websites of public or private organisations

There are some reputable websites which provide useful information relating to Japanese patent litigation in English.

(a)   The IP High Court

The court itself has prepared an English translation or summary for most of the contents on its website (http://www.ip.courts.go.jp/eng/index.html). The website contains an overview of the IP High Court, statistical results relating to IP cases, and guidelines and FAQs relating to lawsuits against trial decision at the JPO. There is also a case database available at the website, although few cases are translated into English.

(b)   The Japan Patent Office

The JPO has a website (http://www.jpo.go.jp/index.htm), and some sections are available in English (eg a section explaining the procedure for obtaining a patent right and a section explaining intellectual property related laws). There is also a database for patent publications, which is searchable in English.

(c)   The Japan Patent Attorneys Association

The Japan Patent Attorneys Association also has a website (http://www.jpaa.or.jp/english/) certain parts of which are available in English. The website also contains information on Japanese patent practice.

---

66   In addition, owing to the size of the task of translating the entire law, English word choices for terms in certain provisions are not necessarily the best for understanding the meaning of the provisions, though not strictly incorrect. For this reason, when a specific provision of the Patent Law is quoted in this chapter, the provision is translated in a way the author believes is best, and it is not necessarily exactly the same as the translation in the aforementioned website.

67   Any person who is trying to gain an understanding of Japanese patent litigation should read the Patent Law. An English translation of the law can be found at the aforementioned website.

4.3   Reliable professionals

For those wishing to gain a better understanding of Japanese patent litigation, the easiest route is to contact a reliable professional. Whilst reading available English information is unlikely to provide a worthwhile understanding of specific issues concerning Japanese patent litigation, a reliable professional can identify the issues and provide the most suitable guidance for each situation.

It is not particularly easy to find such professionals. The internet, however, is a good starting point, as are prestigious law journals or articles such as *Jurist*.

Japan summary

- Japan is a civil law jurisdiction, and has a three-tier court system. The attorney at law acts as an advocate and handles patent litigation with a patent attorney.
- If a patentee finds that his patent has been, or is about to be, infringed by an accused infringer, he should decide the best action after due consideration of many factors including the drawbacks and advantages of each available action (ie regular patent infringement suit; preliminary injunction; import suspension application; or ADR). On the other hand, an accused infringer may pursue a patent invalidation trial or declaratory judgment suit.
- If a patentee files a regular patent infringement suit, the district court holds a series of hearings and divides the proceeding into two stages. The first stage is for infringement and invalidity issues; and the second stage is for damages issues. There is no discovery or concentrated trial. It may take less than a year to reach the decision at district court level. Legal fees and costs are not especially high.
- To prove literal infringement, the claim allegedly infringed should first be examined for ascertaining the technical scope of the patented invention expressed by such claim; and then the construed claim should be applied to the accused product or process. Even if the accused product or process is beyond literal infringement, there is still room for infringement under the doctrine of equivalents. On the other hand, an accused infringer can assert both a non-infringement defence and an invalidity defence.
- There are few reliable sources of information on Japanese patent litigation available in English. Therefore, the easiest and most effective way of gaining some insight is to contact reliable professionals.

I would like to express my thanks for the cooperation of my colleagues, Mr Junichi Kitahara, Mr Hirokazu Honda, Ms Mami Hino, Ms Naho Ebata, and especially Mr Eiji Katayama.

Poland

Paweł Siekierzyński
Gleiss Lutz

Patents for inventions, as well as other industrial property rights, utility models, industrial designs, trademarks, geographical descriptions and integrated circuit systems are regulated in the Industrial Property Law of June 30 2000. This legislation also includes issues relating to registration of industrial property rights by the Polish Patent Office (hereafter the Office; see http://www.uprp.pl/English), registration procedure, litigation procedure before the Office, claims pertaining to industrial property rights and internal organisation of Office (ie functioning and administrative structure). The Industrial Property Law provides that, depending on the particular matter in dispute in a given case, patent-related disputes should be settled either in litigation proceedings before the Office, or in civil litigation before state courts, according to the provisions of the Polish Civil Procedure Code of November 17 1964. The parties to a dispute settled by a decision of the Office can appeal against such a decision to an administrative court.

1. Lawyers

In most cases, the following types of lawyer will be involved in patent cases in Poland:

- *adwokat* (barrister);
- *radca prawny* (legal advisor); and
- patent agents.

Patent agents do not have to have studied law, but should have completed higher studies in a subject which will be of use in their profession as patent agent, in particular technical or legal studies. People belonging to each of the above professions are members of professional associations of *adwokats*, legal advisors and patent agents.

The Polish Civil Procedure Code gives general rights to *adwokats* and legal advisors, to act as legal representatives in all civil cases, in all Polish courts and all court instances, regardless of the location of the court. The Civil Procedure Code also grants the right to patent agents to act in industrial property cases. Patent agents cannot, however, act as legal representatives before the Supreme Court. *Adwokats*, legal advisors and patent agents acting as legal representatives in patent cases have the same powers.

Under the Industrial Property Law of Poland, settling of patent-related disputes before the Office by means of litigation procedure should take place according to the provisions of the Administrative Proceedings Code. This legislation provides that a party to an administrative procedure may empower any natural person to act as an attorney in such procedure. However, in the case of an appeal to an administrative court, the party can usually be represented before the administrative court (as well as before the Supreme Administrative Court) only by an *adwokat*, legal advisor or a patent agent. (It should be noted that, in Poland, the administrative procedure is not a court proceeding and is not a "first instance" within the framework of court proceedings. Under Polish law it is possible to appeal (submit a complaint) against administrative decisions to administrative courts.)

In practice, industrial property cases are usually handled by representatives of all the above-mentioned professions. However, the Industrial Property Law provides exclusive authorisation for patent agents to act before the Office in matters relating to registration and maintenance of protection of industrial property rights.

2. The court system

The Polish civil court system comprises district courts, circuit courts, and courts of appeal. The Supreme Court is a separate judicial body, which is not a regular state court but acts as a final court of appeal. (Again, it should be pointed out that administrative courts are not part of the civil state courts. They constitute a separate element of the judicial system. The administrative courts are also independent of the judicial supervision of the Supreme Court. Appeals against rulings of administrative courts lie with the Supreme Administrative Court.)

According to the provisions of the Civil Procedure Code, district courts do not examine patent matters. Circuit courts are competent in cases relating to copyright protection claims and claims pertaining to inventions, utility models, industrial designs, trademarks, geographical descriptions and integrated circuit systems. Therefore, litigation in patent cases will, in principle, take place before the circuit courts as the court of first instance.

If the circuit court acts as the court of first instance, the bench usually consists of a single judge. The courts of appeal are competent as courts of second instance in cases of appeals against orders issued in the first instance by the circuit courts. The bench of the court of appeals usually comprises three judges.

The Supreme Court is competent to hear cassation claims. Cassation claims are a last-resort appeal, possible only in the event of a violation of the provisions of civil procedure by the court of first and/or second instance, if the violation could have an effect on the verdict or, in the case of a violation of the provisions of substantive law, as a result of misinterpretation or incorrect application of the law. The Supreme Court is also competent to hear claims where the material law on which a final order was based is in issue. The bench of the Supreme Court comprises three judges. The judgments of the Supreme Court are fundamental to the interpretation and application of the law for the state courts. Unanimous judgments of the bench of the Supreme Court, judges from various chambers of the Supreme Court or joint chambers of the Supreme Court will have the effect of a rule of law. Such rulings will also be binding on the Supreme Court. The Supreme Court may either decide a case or refer it back to the lower courts for a decision.

Judges of the circuit courts, courts of appeal and the Supreme Court are highly qualified lawyers, with wide experience in hearing cases and passing judgments. Usually, however, judges do not have a technical education. According to provisions of the Civil Procedure Code, the civil courts are not bound by judgments of any other courts, save for those of the Supreme Court, as mentioned above. Thus, circuit courts are not bound by decisions of the courts of appeal, except in a situation where the court of appeal has returned the case to the court of first instance. In such situations, the legal evaluation of the case and directions regarding further proceedings are binding on the court of first instance, as well as the court of appeal (if the case has returned to the court of appeal in the further proceedings).

The issues relating to patents that the state (civil) courts are competent to hear include:

- ascertainment of the authorship of an invention (ie the author is the inventor);
- ascertainment of the right to a patent;
- setting the level of remuneration for the use of an invention (ie for both non-infringing and infringing use);
- remuneration for the use of an invention for state purposes;
- compensation for the transfer to the State Treasury of a right to a patent for a secret invention;
- patent infringement;
- ascertainment of the right to use an invention in relation to users of an invention, or purchasers of a patent in good faith; and
- the transfer of a patent obtained by a person not entitled to it.

The Industrial Property Law provides that in infringement actions the following claims can be submitted by the holder of a patent or other relevant parties under the law (eg the holder of an exclusive, registered licence):

- cessation of acts threatening infringement of the right;
- cessation of the infringement;
- surrender of unlawfully obtained profits; and
- compensation for damages.

Additionally, the Civil Procedure Code provides that a party, having a legal interest may bring a claim to determine whether a legal relationship or rights exist. This applies in relation to any rights (eg property rights) and any legal relationships (eg agreements).

The Industrial Property Law states that claims can be submitted according to the civil procedure regulations, before the state courts. In practice, however, a party that claims its rights have been infringed, before taking its claim to court, will often issue a warning letter and demand that its opponent cease the infringement. This may result in negotiations, and in many cases the parties reach a settlement as a result of such negotiations, rather than engaging in litigation.

2.1 Arbitration

Under the Civil Procedure Code, the parties can submit a dispute (ie either an existing or a possible future dispute) relating to a patent to the arbitration court, on the basis of a written agreement concluded between the parties to the dispute. The parties may determine in the agreement the number of the arbiters and how they are appointed. If this matter is not determined by the parties then, under the Civil Procedure Code, the arbitration court will be composed of three arbiters. The Civil Procedure Code will also apply with regard to the method of appointment. The ruling of the arbitration court (or an agreement concluded before it) is equivalent to the ruling of a state court (or an agreement concluded before the state court) once the state court has recognised the ruling, or has stated that the ruling is enforceable. Under the Civil Procedure Code, if the dispute is submitted to arbitration, the claimant can still apply to the state court for protective measures (interim relief), in order to secure the claim.

2.2 Competence of the Polish Patent Office

Under the Industrial Property Law, the Office is exclusively competent to settle certain litigious industrial property cases. These cases can be classed as 'revocation actions'. This includes the following:

- invalidation of a patent;
- invalidation of a European Patent, granted according to the provisions of the Convention on the European Patent;
- expiry of a patent for an invention concerning biological material, or for use, in the case of the permanent loss of possibility of using the invention due to lack of the appropriate biological material required; and
- invalidation of a patent, in consequence of an opposition filed, which is claimed by the rights holder to be unjustified. (Although Polish legal terminology does not include the term 'contested opposition', this is similar to contested opposition.)

The Office resolves the above matters in litigation proceedings. It is a quasi court procedure, regulated by the provisions of the Industrial Property Law and of the Administrative Proceedings Code. Litigation proceedings should be completed within six months from the submission of the request for initiation of a proceeding to the Office.

Cases brought in the Office are heard by an adjudication panel, composed, in principle, of three persons: the chairman, and two members of the panel. The chairman of the panel is an employee of the Office, having the necessary legal training and extensive professional experience. The other members of the panel are two experts, authorised by the President of the Office to adjudicate in litigation proceedings.

According to the Industrial Property Law, the only people who may be admitted as experts are those who hold a university degree and have graduated from a relevant faculty, who have completed the necessary practical training for experts, and who have served their apprenticeship as an assistant expert.

If a party to litigation proceedings deems that a decision made, or an order issued, in the litigation proceedings violates the law, the provisions of the Industrial Property Law and the Act on Administrative Court Proceedings of August 30 2002 provide that the party may issue a complaint to the administrative court within 30 days of the date of the decision, through the Office. (The complaint to the administrative court is, in effect, an appeal and the administrative courts are a separate class of courts.) The Office should, within 30 days, deliver the complaint and the files of the case to the administrative court, and it may allow the complaint within this period. Lodging the complaint does not automatically suspend the execution of the decision or the order; however, the Office or the administrative court may suspend it.

A party to the litigation proceedings which has not lodged an appeal is automatically a party to the proceeding before the administrative court if the outcome of the court proceedings is connected with his legal interest.

Under the Act on Administrative Court Proceedings, the court's bench will usually comprise three judges.

In the event of a violation of the substantive law or procedure, the parties have a right to cassation within 30 days, in the Supreme Administrative Court. (The Supreme Administrative Court is the court of second instance for verdicts of the administrative courts. The administrative courts system was established to provide for judicial verification of decisions of administrative bodies.)

The court will usually consist of three judges.

The provisions of the Civil Procedure Code and those relating to proceedings before the Office, do not generally stipulate any specific requirements that the parties must satisfy before filing the statement of claim at court, or the application at the Office; the claimant/applicant need only pay the required fee. The Civil Procedure Code provides specific requirements, however, in the case of disputes between entrepreneurs. (An entrepreneur is any person, partnership, company or legal entity that has been registered in the respective Entrepreneurs' Registers.) According to these provisions, before filing a statement of claim at the court, a claimant must issue a warning letter to the infringing party. Moreover, the claimant is obliged to present all his statements and evidence to support them in the statement of claim.

3. Procedure and timescale of proceedings

3.1 Filing infringement actions

In the case of patent actions filed in accordance with civil procedure before the state court, the procedure commences when a statement of claim is filed at court by the claimant. (The Civil Procedure Code contains regulations regarding the territorial jurisdiction of the courts. The basic principle is that the domicile of the defendant determines the court at which the statement of claim should be filed. There is no particular circuit/appeal court which would have significantly more experience than other courts.) The statement of claim should include a detailed description of the claim and, in cases concerning property claims, the claimant should state the value of the litigation claim, set out the circumstances justifying the claim, and the circumstances justifying the competence of the court (if necessary).

In the case of disputes between entrepreneurs, the specific provisions of the Civil Procedure Code will be applicable, concerning the proceedings in business cases. According to these provisions, the claimant must present all his statements and evidence to support them in the statement of claim, because the court will not admit them at a later stage in the proceedings. The claimant may, however, disclose the statements and evidence later, if he can prove that it was not possible to include them in the statement of claim or that it became necessary to disclose them at a later date. The claimant should provide the court with both the statement of claim and a copy of a warning letter containing a demand for voluntary fulfilment of the claim, or confirming that the claimant has attempted to resolve the dispute by means of negotiations.

Under the general provisions of the Civil Procedure Code, once the statement of claim is filed at court, the court examines it to ensure that it meets all formal requirements, according to the provisions of the Civil Procedure Code, and to determine whether the correct fee for filing it at court has been paid. If the court finds the statement of claim is defective as to legal formalities, it requests the claimant to correct it within one week; otherwise it returns the statement of claim.

Regardless of the type of the proceedings (ie general or business cases), the court always examines whether it has competence to hear the case, sets the date of the hearing, and orders service of the statement of claim and a demand to appear at the hearing on the defendant. The defendant may file his reply to the statement of claim (defence) with the court. (Business cases are cases which involve entrepreneurs, and concern their business activity. A patent case is a business case only if it concerns a dispute between entrepreneurs.)

Should the court deem that the case is complicated or involves financial settlement between the parties, it may order that the defendant replies to the statement of claim before the first hearing. It can also order the parties to exchange further preparatory letters in order to accelerate the proceedings. In business proceedings, the defendant is obliged to file the defence to the statement of claim at court.

If the party is represented by an *adwokat*, legal advisor or a patent agent, the court may place it under an obligation to submit within a specific deadline a preparatory letter in which the party should disclose all its statements, claims, objections and evidence, under pain of loss of right to disclose them later in the proceedings.

The hearing is adversarial and starts with an announcement of the case in court. The parties then set out their positions. The claimant and the defendant may propose their demands and motions and present the statements and evidence to support them. The court should establish which of the relevant circumstances of the case are in dispute and aim to resolve them. If the case is sufficiently clear, the court can make a ruling. It can also dismiss the action and discontinue the case. Once the parties present their positions, an agreement can be reached in court. If the litigation is not fully resolved at this stage of the hearing, and relevant factual circumstances have not been clarified, the hearing of evidence is conducted.

The court closes the hearing if it determines that the case has been sufficiently clarified. In the general procedure, the parties may present the factual circumstances and evidence at any stage before the hearing is closed. Before closure of the hearing, the parties may present their closing statements and summarise their cases; and after the hearing is closed, the court makes a ruling.

In proceedings in business cases, there are a number of differences. For example, during the proceeding, neither party may submit new claims in addition to/instead of the previous claims, and the party represented by an *adwokat*, legal advisor or patent agent should deliver the litigation documents containing the parties' statements directly to the opposing party. At the request of one of the parties, the court may order that the hearing will be a closed hearing, if confidential information or circumstances may be disclosed. The Civil Procedure Code provides that the court should aim to make a ruling within three months of the date of filing the statement of claim. The Civil Procedure Code does not have any form of accelerated proceedings in industrial property cases. In terms of timing, the aforementioned three-month period relates only to proceedings in business cases. "General" proceedings do not have such a timeframe. The three-month period starts with filing the statement of claim at the relevant state court. The actual period of proceedings depends on the number of cases in a given court.

The regulations of the Civil Procedure Code provide that the proceeding may be suspended if the settlement of the case depends on the outcome of other civil litigation or a previous decision of a public administrative body. This includes revocation actions in the Office and any appeal to the administrative court. However, the civil court cannot stay proceedings in relation to an action before the European Patent Office.

3.2 Filing revocation actions

In litigation proceedings before the adjudication panel of the Office, litigation cases are resolved in administrative hearings. Before the hearing, all necessary steps are taken to enable the proceedings to be conducted. In particular, the parties are required by the Office to submit explanations, documents and other evidence. Additionally, the parties must attend the hearing in person, or through their legal representatives or holders of a power of attorney authorising representation. Witnesses and experts are also required to attend the hearing. If the parties have been summoned to attend and fail to do so, this will not prevent the hearing taking place. At the hearing the parties may submit clarification, put forward their demands, proposals and complaints and provide evidence in support. The parties may also state their views on the outcome of the procedure relating to evidence (ie they may present their opinion as to the conclusions of inspections and what has been stated by the court and the other party about the evidence presented).

Under the Administrative Proceedings Code, litigation proceedings may be suspended. This is also the case if a decision is dependent on prior resolution of a preliminary issue by another body or court.

The Office has the competence to examine matters on annulment of a European Patent. This arises from the procedure set out in the European Patents Convention;

however, as yet the Office has not examined this type of matter, and cannot rely on previous practice relating to the possibility of suspension of proceedings due to a proceeding before the European Patent Office. As regards Polish patents, proceedings in progress before the European Patent Office have no direct influence on litigation proceedings and, in particular, cannot be grounds on which to suspend litigation proceedings.

If an appeal is filed at the administrative court, the court passes judgment at the hearing. On a party's application, the court orders a closed hearing if this is necessary to safeguard that party's personal life or its private interests.

Once it is announced in court, the case commences with a statement by the judge, briefly setting out the matter based on the case files, and in particular taking into account the allegations made in the appeal.

Next, the parties – first the appellant and then the Office – orally state their demands and proposals and set out their reasons.

4. Mediation

In the case of infringement actions, the Civil Procedure Code provides for the possibility of mediation between the parties before the civil litigation commences. The parties may either conclude an agreement, according to which they attempt to resolve certain disputes by mediation, or the court may order them to attempt a resolution by mediation. However, the mediation is voluntary. Any settlement reached before the mediator has the same legal effect as a court settlement. Should one party rely in the litigation or arbitration on the settlement proposals, or any proposal for a compromise settlement, or other statements submitted in the mediation, such reliance will be ineffective.

5. Availability of pre-action evidence gathering

There is no automatic discovery or disclosure of documents in patent proceedings in Poland. Likewise, there is no mechanism for obtaining documents or other evidence from a potential defendant other than where the defendant volunteers such material (eg during mediation). However, before the hearing, the court may order, according to the statement of claim and other litigation documents containing the parties' statements, that documents, objects to be inspected, books, plans and so on be produced. Further, as described below, a patentee may apply for the provision of certain information.

6. Availability of interim relief

The provisions of Poland's Industrial Property Law have been harmonised with those of the European Enforcement Directive 2004/48/EC and, as a result, interim relief for protection of evidence has been introduced into the Industrial Property Law. An application for this relief can be made either before the statement of claim is filed at court or at a later stage.

With regard to protective interim relief, the provisions of the Civil Procedure Code (unlike those of Directive 2004/48/EC), do not contain a specific list of protective interim measures that can be ordered by the courts in the case of non-monetary patent claims. According to the provisions of the Civil Procedure Code, even if the subject of the relief is not a monetary claim, the court can at its discretion grant the protective relief typical for monetary claims, taking into account the given circumstances. In particular, the court can:

- regulate the rights and obligations of the parties for the period of the litigation;
- prohibit the transfer of the objects or rights which are part of the litigation claim;
- stay the enforcement; and
- order registration of a caution in the land and mortgage register, or in any other appropriate register.

The court serves the party which has been placed under an obligation with the interim relief order made in a hearing in chambers. The relief will impose an obligation on the party to carry out a positive act, or prohibit it from doing something, or to cease interfering with the rights of the patent holder. Therefore, the court may apply the measures set out in article 9 of the Enforcement Directive. The order will be granted only if certain conditions are fulfilled:

- the patentee presents his/her reasons for requesting interim relief,
- he/she shows that the circumstances alleged in the claim are probable; and
- he/she demonstrates the urgency of the matter.

The court grants the interim relief only at the application of the claimant, and the defendant does not need to be present at the hearing in chambers.

Under the Civil Procedure Code, the court may make the protective interim relief conditional on the patent holder submitting a deposit as a security for any claims of the defendant which may arise as a result of any damage caused by enforcement of the interim relief.

The court may issue an interim relief order before or during the hearing. The patent holder, requesting the protective relief, should state his legal interest (eg the circumstances which may result in damage), unless the court grants provisional and protective measures. The court considers the application for protective relief at a hearing in chambers. The provisions of the Civil Procedure Code, which correspond to the provisions of the Enforcement Directive, provide that the defendant may file a complaint against the protective relief. The defendant may also apply to revoke or change the protective relief. The court should make its decision after the hearing. If the court makes an order for protection of evidence or for protective relief before infringement proceedings have been commenced, it should also set a time period (maximum two weeks) within which the patent holder should file the court document initiating the proceedings at court. If the patent holder fails to submit it within that period, the protective relief will be ineffective.

The Civil Procedure Code also enables the court to place the defendant under an obligation to make a single payment, or periodic penalty payments, if the patent holder demonstrates that his claim is credible. These can be imposed before or during litigation at the request of the patentee. The conditions for granting the order are the same as those for granting other interim relief. The payments go to the patentee.

The protective relief will become ineffective if the statement of claim or application is returned or dismissed, or the proceedings are discontinued; and if the protective relief was given before the commencement of the proceedings, it will also become ineffective if the patent holder does not demand in the proceedings satisfaction of the entire claim or has filed a different claim, instead of the claim that was protected by the court's order.

Under the Industrial Property Law, a patent holder can file, at a court that is competent to adjudicate on industrial property cases (ie a court which has territorial jurisdiction for the location of the registered office of the infringing party or place of domicile), an application for an injunction obliging the person who infringes the patent, or a third party, to provide any information which can be helpful in demanding cessation of the infringement, surrender of unlawfully obtained profits and compensation for damages. All relevant information should be delivered before the main hearing to the competent state court. The court evaluates the necessity for issuing an injunction at its own discretion, concerning the origin, and the retail network of the goods or services that infringe the patent, if the infringement is highly likely to occur, and if such actions are aimed to achieve economic benefit. The aforesaid information may concern:

- the names and addresses of the manufacturers, retailers, suppliers and other previous holders of the goods, or service providers, as well as recipients of the goods or services; and
- the quantity of the goods or services manufactured, sold, received or ordered which infringe the patent, and the prices paid.

7. Evidence

7.1 Civil litigation

The hearing of the evidence takes place before the court which conducts the litigation. According to the provisions of the Civil Procedure Code, the evidence consists of facts relevant to the settlement of the dispute. Evidence need not be produced in support of facts which are apparent and universally accepted. Facts which have been acknowledged during the proceeding by the opposing party do not need to be proved if the acknowledgement is irrefutable. The parties are obliged to adduce evidence in support of the facts, from which the parties derive legal consequences. The court may allow material evidence, in particular films, photographs, plans, drawings or tapes. If a party does not expressly disagree with statements of the other party concerning the facts, the court may deem those facts as acknowledged by that party.

The civil court evaluates the credibility of the evidence, taking into consideration the material gathered as a whole. The court is bound by legal presumptions (ie presumptions established by the law in a number of legal acts, for example the presumption that possession is consistent with the state of the law and that the possessor is the owner); however they can be refuted. The court can alter or overrule its decision regarding the evidence at any time. Failure by a party to appear at court on the hearing date does not prevent the evidence from being heard, unless the presence of a party (or parties) is necessary.

Official documents (eg a decision of the Office concerning granting a patent for an invention), drawn up in the appropriate form by the relevant public authorities within the scope of their competence, are proof that their contents have been officially certified. A statement in a document, which is not an official document, is taken to be that of the person who has signed the document.

According to the provisions of the Civil Procedure Code, if an act or an agreement concluded between the parties is required to be in writing, the parties to the agreement or participants in the legal act can prove the conclusion or performance of the agreement only if the document concerning this legal act was lost, destroyed or taken by a third party.

Where a party challenges the authenticity of an official document or claims that the statements of a public authority are untrue, or where it is claimed that a document which is not an official document is untrue, or that the statement in the document is not that of the person who signed it, evidence must be provided to support such claims. If a party challenges the authenticity of a document which is not an official document (ie not a document issued by public authorities) or maintains that the statement in it is not that of the person signing the document, the challenging party must prove its allegations. If, however, the dispute relates to a document which is not an official document which has been drawn up by someone other than the party making the allegation, the party which intends to use the document as evidence must prove that it is genuine.

If a party calls witnesses, it should specifically indicate facts that should be proven by the testimony of each witness. It should also indicate the witnesses, in order for the court to call them. The Civil Procedure Code provides that if the witnesses offer conflicting testimonies, the witnesses may be challenged. Thus, during each hearing of a witness, each party can examine the witness and ask him/her questions.

If specialist knowledge is required, the court may call one or several experts. The court decides whether the experts' opinion should be presented in writing, or orally. The court may require an oral explanation of a written opinion, and can request an additional opinion from the same or other experts. The court may also request an opinion from a scientific institute or scientific research institute.

If the evidence has been exhausted, or no evidence has been produced, and facts relevant to the settlement remain unclear, the court will order the hearing of the parties in order to clarify those facts.

7.2 Proceedings before the Office and the administrative courts

In litigation proceedings before the adjudication panel of the Office, the evidence is heard in accordance with the provisions of the Administrative Proceedings Code. Anything which may assist in clarifying the matter will be allowed as evidence if it is not contrary to the law. Evidence will usually be provided in the form of documents, witness testimonies and expert opinions. Proceedings are conducted in a similar manner to the hearing of evidence in civil proceedings, but with certain exceptions (eg the provisions of the Administrative Proceedings Code do not allow for the possibility of requesting a scientific institute or a research and development centre for an opinion). However, according to rulings of the Supreme Administrative Court this type of evidence is allowed.

The administrative court may *ex officio*, or on the application of one of the parties, consider documentary evidence if this is necessary to clarify serious doubts and does not lengthen the proceedings unduly. The court takes facts which are apparent and universally accepted into consideration, even if the parties do not rely on them. Moreover, if the parties or their representatives are not present at the hearing, this does not prevent the hearing from taking place.

8. Law

A patent infringement occurs if one person encroaches on the exclusive rights of the patent holder for its own profit. The doctrine of industrial property law, as a rule, accepts that the infringement of a patent takes place only if in the industrial activity the whole technical solution designated in the patent claim was used. Therefore, there will be no infringement of a patent if at least one of the characteristics claimed is missing.

Polish law does not contain legal guidelines regarding the interpretation of patent claims. According to opinions presented in the doctrine of industrial property law, the court should put the emphasis on the literal interpretation of the claims and not deviate significantly from this. The doctrine of equivalence is used mostly in the assessment of the patentability of an invention. There are no procedural rules in terms of determining the scope of previously granted patents.

8.1 The patent

Under the Industrial Property Law, patents are granted, regardless of the field of technology, for inventions which are new, which involve an inventive step, and which are suitable for industrial application. An invention is deemed to be new if it is not a part of the state of the technology. Under the Industrial Property Law "state of the technology" means everything which prior to the date of applying for the registration of the patent was published for the general public anywhere in the world. An invention will be considered as involving an inventive step if, having regard to the state of the technology, it is not obvious to a person skilled in the technology. It is possible to grant a patent for an invention concerning a new use of a substance comprised in the state of the technology, or the use of such substance for the purpose of obtaining a product for a new use.

An invention is considered as capable of industrial application if by means of that invention a product may be produced or a process may be used, in a technical sense, in any kind of industry, including in agriculture.

Current doctrine takes the view, which is common practice in Supreme Court rulings, that in order to demonstrate that a given solution is not inventive, it must be shown that there are solutions in the state of the technology which in essence have the same group of qualities. According to doctrine and Supreme Court rulings, in order to assess the degree of inventiveness (which is not obvious), one must compare the group of technical characteristics of the invention with the closest solution in the state of the technology. There are a number of circumstances that, if present, mean that a given solution satisfies the condition of not being obvious. These include solutions to a problem which has been unsuccessfully tackled by experts in a given area, which gives a definite economic result, or which is satisfying a new social need. The effect of these circumstances depends on the circumstances of a given case and will be determined by the Office.

Doctrine also indicates that using a solution of a given scope, which is covered by a patent, does not exclude the granting of another patent for a solution included in a previous patent, provided it satisfies the prerequisites of novelty and the fact that it is not obvious. The subject of examining the degree of the inventive step and the patent claim is varied; however, the fact that a solution which is contradictory to an earlier solution was itself patented does not prevent a finding that infringement of the patent has taken place.

According to rulings of the Supreme Court, a difference in technical means in two inventive solutions does not allow a conclusion to be drawn that one of these solutions is an obvious consequence of the other. Furthermore, the court has ruled that even if only one substance is changed, causing significant changes as to the quantity or quality of the product, it causes an inventive step. This is so, even if the change was obvious, because neither the provisions of the law nor jurisprudence stipulate that the inventive step should be unexpected.

According to provisions of the Industrial Property Law, a biotechnological invention means an invention, within the aforesaid meaning, concerning a product consisting of or containing biological material, or a process by means of which biological material is produced, processed or used. Moreover, under Poland's Industrial Property Law, the following are considered as biological inventions eligible for patent protection:

- inventions, the subject of which is biological material which is isolated from its natural environment or produced by means of a technical process, even if it was present in the natural environment;
- elements isolated from the human body or otherwise produced by means of a technical process, including the sequence or partial sequence of a gene, even if the structure of that element is identical to that of a natural element; and
- inventions which concern plants or animals, if the application of the invention is not confined to a particular plant or animal variety.

A sequence or a partial sequence of a gene can only be the subject of a patent if it has an industrial application. The industrial application must be disclosed in the patent application.

The Industrial Property Law provides that the human body, at the various stages of its formation and development, and the mere discovery of one of its elements, including the sequence or partial sequence of a gene, cannot constitute patentable inventions. Moreover, the Industrial Property Law provides that processes for cloning human beings, processes for modifying the genetic engineering of embryonic stem cells of human beings, the use of human embryos for industrial or commercial purposes, processes for modifying the genetic identity of animals which are likely to cause them suffering without any substantial medical benefit to man or animal, and also animals resulting from such processes, are considered as biotechnological inventions whose exploitation is contrary to public order or public morality. Therefore, patents for such things are not permitted.

Protection conferred by a patent on a biological material possessing specific characteristics indicated in a patent claim as a result of the invention extends to any biological material derived from that biological material through propagation or multiplication in an identical or divergent form and possessing those same characteristics. The protection conferred by a patent on a process that enables a biological material to be produced, possessing specific characteristics indicated in a patent claim as a result of the invention, will extend to biological material directly obtained through that process and to any other biological material derived from the directly obtained biological material through propagation or multiplication in an identical or divergent form and possessing those same characteristics.

9. Available remedies

The provisions of Polish law, relating to industrial property claims, have been harmonised with the provisions of the EU Enforcement Directive 2004/48/EC.

Subject to the provisions of the Industrial Property Law, claims for infringement of a patent are possible only after the patent has been granted. However, where the infringing party has acted in good faith (ie without knowledge of the patent application), claims for the infringement of a patent can be made in respect of the period beginning on the day following the date of publication by the Office of the patent application. If the infringing person has been notified in advance, by the holder, of the filing of the patent application, they are liable for infringement from the date of notice.

The statutory limitation period for claims for infringement of the patent is three years. This period runs, separately in respect of each individual infringement, from the date on which the holder of the right learned about the infringement of his patent and about the infringing person. However, in any case, the claim becomes statute barred five years after the date on which the infringement has occurred.

When ruling on the infringement of a patent, the court may, at the owner's request, decide as to unlawfully manufactured or marked products and on the means used in their manufacturing or marking, which are owned by the infringing person. In particular, the court may order their withdrawal from the market, or award them to the patent holder in order to count them in any calculation of damages, or order their destruction.

A patent holder whose patent has been infringed, or any person enjoying the same status, may demand from the infringing party cessation of the infringement, surrender of unlawfully obtained profits and, if the infringement was culpable (ie carried out knowingly), compensation in the form of damages. Damages may be sought in accordance with the general principles of law, or by means of payment amounting to the equivalent of the licence fee, royalties or other fees which would have been due if the infringer had requested authorisation to use the patent in question. In either case, damages will include losses incurred by the patent holder, as well as benefits which he could have obtained had he not suffered the damage.

If the level of damages, according to the general principles of law, is difficult or impossible to determine, the civil court may award in the judgment an appropriate amount at its discretion. At the request of the patent holder, the court may rule that the judgment be published in part or in its entirety. (Generally, judgments are not made publicly available unless requested by a party. Moreover, Supreme Court hearings are not open to the general public unless certain circumstances pertain, for example where the matter at hand contains a relevant legal issue, or the party which has submitted the cassation complaint (last-resort appeal) requested a public hearing.) If the infringement was not carried out knowingly, the court may order the infringing person, at his request, to pay an appropriate amount to the patent holder, in circumstances where the cessation of the infringement or forfeiture or destruction of the goods would cause severe hardship to the infringing person and this would not be detrimental to the interests of the patent holder.

According to the provisions of the Civil Procedure Code, the judgment is in principle enforceable once it becomes legally valid (ie if the provisions of the civil procedure provide no possibility of an appeal to a higher instance). The Civil Procedure Code includes regulations which allow the court *ex officio* to order immediate enforcement of the court's order in a number of situations. These include orders for a claim accepted by the defendant, or default judgments complying with the statement of claim. Upon a party's application, the court may rule immediate enforcement of a judgment if a delay would prevent or significantly hinder the enforcement of the judgment, or may result in damage. Subject to the regulations of the Civil Procedure Code, the court will not rule immediate enforcement if this could cause irreparable harm to the defendant. The court may make the immediate enforcement conditional on the claimant submitting appropriate security.

Should the immediately enforceable judgment be revoked or changed, the court (on the defendant's application) will make a ruling regarding the restoration or return of the executed performance.

10. Costs

Under the provisions of the Polish Civil Procedure Code, the losing party, at the request of the other party, should refund the necessary costs incurred. The necessary costs will usually include the fees of one *adwokat*, legal advisor or patent agent, their related expenses, court fees, as well as costs of mandatory personal appearances of the party ordered by the court. The fees of a professional legal representative cannot exceed the amounts indicated in the relevant provisions.

If the court resolves the claim partly in favour of one party and partly in favour of the other, the costs are set off against each other, or divided respectively. The court may, however, rule that one party will be obliged to return all costs if the opposing party has been unsuccessful in a major part of its claim. Moreover, the defendant is entitled to a refund of the costs, even if the claim was settled, if he did not give grounds for litigation and acknowledged the claim at the first opportunity in the proceeding (ie the court determines whether the losing party gave grounds for litigation, and whether the other party prevailed decisively or insignificantly). Based on this determination, the court can award the costs proportionally between the parties.

Court fees in civil cases are regulated in the Act of July 28 2005, on Court Costs in Civil Law Cases.

The costs that can be incurred in patent litigation before the civil court, depend in principle on the nature of the claim (ie if the claim is a non-property claim or a property claim) and on the nature of the parties' written requests submitted to the court. In the event of an application for interim relief for protection of evidence (as well as other applications for protective relief, unless they were filed at court together with the statement of claim), the court fee is PLN40 (PLN100, in the case of property claims).

The fee for filing statements of claim in patent cases in non-property claims is PLN600. This fee is fixed, regardless of the value of the litigation claim. For filing a statement of claim in property claims, the court fee is 5% of the value of the litigation claim, but not less than PLN30 and no more than PLN100,000.

As regards the costs of the proceedings before the Office, the above provisions relating to civil proceedings apply. The secondary provisions (ie executory provisions to the Industrial Property Law) provide for a fee for the application for a decision in litigation proceedings amounting to PLN1,000. The Administrative Proceedings Code provides that a party bears the costs which arise through its fault or which were incurred in its interests, and do not arise from the statutory obligation of the Office.

The above provisions also provide for a fee of PLN1,000 on the application for an appeal to the administrative court. If the court of first instance upholds the appeal, the appellant has the right to have the costs of proceedings refunded by the Office which made the order, or which undertook actions or failed to act, and on the basis of which he has therefore appealed. A party represented by an *adwokat*, legal advisor or patent agent includes their fees in his costs. However, this figure cannot be higher than the rates defined in separate provisions and expenses of one *adwokat*, legal advisor or patent agent, court costs and costs of court attendance of a party which was ordered by the court.

11.   Hot topics

11.1   Recent changes in the law

In the second half of 2007, new legislation in Poland set out provisions relating to protection of patent rights. These changes result from the implementation of Directive 2004/48/EC of the European Parliament and Council dated April 29 2004 into Polish law. These provisions relate to enforcement of intellectual property rights and in particular to the provisions of the Industrial Property Law.

The Amending Act of the Industrial Property Law relates, among other things, to court rulings in matters relating to goods produced as a result of the infringement of patents. Under the previous regulations the court had general competence, when an application was made by a patent holder, on the disposal over goods which were unlawfully produced or marked and the measures used for this purpose. This provision was extended, and the current remedies are described above. Moreover, in implementing Directive 2004/48/EC, the legislation introduces into the Industrial Property Law interim relief for protection of evidence and for security of a claim. Both protective measures have been discussed above.

This regulation creates new possibilities for the protection of parties with rights under a patent and considerably simplifies claims arising from infringement of patents. The amendment of the provision makes it possible for the patent holder to bring a claim for compensation for damages more effectively. The regulations which were previously in force only provided for compensation for damages on general principles, which in practice caused problems in calculating the damage suffered. The amended provision introduces additional possibilities for compensation, as described above.

These changes in the provisions considerably simplify a patent holder's protection of its rights, as well as claims for cessation of infringements.

Poland summary

- Depending of the type of claim in hand, cases relating to patent rights can be resolved either in litigation proceedings before the Office, as defined in the Industrial Property Law and the Administrative Proceedings Code (and in the case of appeals to the administrative court, in accordance with the Act on Administrative Court Proceedings), or civil proceedings before the state courts.
- Patent revocation proceedings are filed in the Polish Patent Office and should end within six months of the commencement date. The party which intends to initiate the proceedings should file an application to the Office and pay the fee of PLN1,000.
- If an appeal is filed at the administrative court, the person filing the appeal must pay a fee of PLN1,000. The legal provisions do not state the date by which the court or administration proceedings must end.
- For matters resolved in civil proceedings (infringement and so on) before the state courts, the general provisions do not define a time period within which a matter must be resolved; however, the provisions relating to proceedings in economic matters state that the court should attempt to make a ruling within three months of filing the statement of claim.
- If a claim is filed for the protection of non-property rights arising from the patent, the fee is PLN600. In the case of property claims, the court fee is 5% of the value of the claim but in any event not less than PLN30 and not more than PLN100,000.

Portugal

Vasco Stilwell d'Andrade
Morais Leitão, Galvão Teles, Soares da Silva & Associados
Gonçalo da Cunha Ferreira
Garrigues

Over the last few years, the Portuguese government has been deeply committed to transforming Portugal's economy into one based on vanguard technologies. According to the European Innovation Scoreboard, Portugal now ranks seventh among the economies that have progressed most in terms of innovation.[1] Consequently, much emphasis has been placed on the importance of intellectual property, and the country has seen an unprecedented number of public and privately funded initiatives promoting the benefits of intellectual property (IP) protection.

As a result of this greater awareness of intellectual property, the number of national patent applications has witnessed a slight increase,[2] as has the number of patent litigation cases. Nevertheless, the truth remains that Portugal continues to be one of the countries with the lowest number of patent applications in relation to gross domestic product, and it has fallen behind countries like Brazil and China. Likewise, patent litigation is somewhat rare in Portugal and very few cases reach the courts each year. The vast majority of patent-related disputes are resolved through extrajudicial agreements, leaving Portuguese lawyers, judges and legal scholars with very little in the way of case law.

This chapter will seek to provide an objective, realistic and comprehensive picture of patent litigation in Portugal today, with all its virtues and flaws. It should be noted that the information provided is presented in a very broad manner and conveys only the general rules and perspectives that are currently applicable in this country. Portuguese procedural law is complex and there are numerous exceptions to the general rules. Explaining all the intricacies of Portuguese procedural law clearly exceeds the scope of this chapter. Additionally, one must point out that legislative changes and reforms are extremely frequent in Portugal. It is therefore of the utmost importance to obtain a detailed and updated analysis of each case from a local law firm before deciding to move forward with patent litigation in Portugal.

1. Lawyers

It is obligatory to retain at least one lawyer for all patent litigation proceedings in civil or criminal judicial courts. Portugal is a small country operating under a single unified judicial system and therefore any lawyer can represent the client in any court

---

1   www.proinno-europe.eu
2   In 2006, 88 European patent applications were filed by Portuguese applicants. This was the highest number ever and up from 50 in 2004 and 66 in 2005 (*Source: Jornal "Expresso"* February 16 2008).

of the country.[3] Following EU directives transposed into internal legislation (in particular Directive 98/5/EC), foreign lawyers may litigate in Portuguese courts when acting under guidance, and with the assistance, of a lawyer accredited by the Portuguese Bar Association or when they themselves have been accredited by that same association. Voluntary arbitration courts and other extrajudicial dispute resolution systems have more flexible rules and much will depend on the terms of the arbitration.

Given the technical nature of patents, which typically cause some unease among lawyers, it is usual to obtain the assistance of a technical expert for both the initial analysis and during trial. Portuguese procedural law specifically permits that a lawyer be assisted during the presentation and discussion of evidence by a technical expert. It is, however, necessary for the lawyer to give ten days' prior notice of who will be assisting him and what issues require technical explanation. The other party will also have a right to do so. The court may deny the appointment of technical experts if it feels that the matter at hand does not justify their participation.

As regards technical experts, it is worth pointing out that in Portugal there is a group of professionals who are recognised by the Portuguese Patent and Trademark Office (PTO) as representatives with an advanced degree of knowledge in IP-related issues. They are known as AOPIs or *Agentes Oficiais da Propriedade Industrial*, which directly translates as official industrial property agents. Whilst AOPIs are very knowledgeable on IP matters, they do not necessarily possess a technical or scientific background like European patent attorneys and are often not experts in a particular area of technology. For very detailed and complex technical matters in court it is therefore necessary to resort to university professors, experts with PhDs or other recognised technical experts in a certain field. Nevertheless, all other things being equal, it is probably wise to use the services of a lawyer who is also an AOPI, or someone with recognised experience in patent matters. Patent litigation teams in Portugal that actually defend a case in court rarely extend beyond one or two people, although in the back office there are often various other lawyers analysing and preparing the case.

2.   Law

The Portuguese Republic is a founding contracting party to the 1883 Paris Convention and its revisions, and a member of the Patent Cooperation Treaty (PCT), the European Patent Convention (EPC), the Agreement on Trade Related Aspects of Intellectual Property Rights (TRIPS) and most other international agreements on patent-related matters.[4] It has further implemented EU directives into internal law and is an eager member of all patent discussion forums.

It is therefore no surprise that Portuguese substantive patent law is today closely harmonised with the EPC and, in relation to the more controversial aspects

---

3   Despite being a small country, Portugal has two sets of island archipelagos in the middle of the Atlantic (Madeira and Azores), which constitute two autonomous regions. Though a lawyer from the continent can technically represent the client in the courts of these islands, there will be additional costs in terms of travel expenses.

4   Portugal is a signatory to the Patent Law Treaty (2000) but at the time of writing had not yet ratified it.

(patentability of software, business methods, biotech inventions), it follows the positions held by the European Patent Office (EPO). Those who wish to proceed with patent litigation in Portugal can expect to find essentially the same substantive rules as those found in other European jurisdictions and the EPC.[5] Given the limited domestic case law available on substantive patent law, it is customary to resort to EPO decisions and guidelines in the interpretation of concepts such as novelty, inventive step and on exceptions to patentability. Overall, too few cases have been taken to higher-instance courts to establish any sort of coherent jurisprudence in most patent matters. Issues such as whether the doctrine of equivalence is admissible or whether the literal test is preferred have not been adequately discussed in national courts, although there is case law[6] in Portugal that clearly leans towards the acceptance of equivalence.[7] One should bear in mind that even if more case law were available, Portugal does not follow the rule of precedence and every judge is free to decide as he sees fit, even if it goes against rulings of the Supreme Court.[8] This general lack of jurisprudence makes local lawyers rely heavily on domestic academic law expert opinions and examples from abroad on questions of law.

Domestic patent application procedural rules up to grant also follow the general outline of those established in the European Patent Convention in terms of timings and requirements. A search report is issued during the eighth month after filing, and the application is published during the eighteenth month from the filing date or the claimed priority date. One major difference in relation to the European Patent procedure is that the opposition period begins immediately after the publication at 18 months and not after an intention to grant. Opposition must be filed within two months of the application publication and the applicant then has a similar deadline in which to reply. Filing additional replies is possible under special circumstances and subject to the Portuguese PTO's approval. In Portugal, the PTO performs a substantive examination on patentability and the final decision to grant or refuse the application will be based on the conclusions of that examination and any other evidence and arguments presented during opposition. Even after the issuance of a final decision to grant, it is still possible to request its revision by a superior official within the PTO or appeal it in court.[9]

Another particularity of Portuguese substantive patent law worthy of note is that in order to establish a priority date in Portugal, it is not necessary to file a complete application with claims, description, drawings and abstract. The recently reformed 2003 Portuguese Industrial Property Code (PIPC) follows the approach of article 5, paragraph 1(iii) of the Patent Law Treaty (Geneva, 2000), which requires that only an

---

5   It should be noted, however, that up to 1995 it was not possible to patent pharmaceutical or chemical products, as such. This has led to a great deal of patent litigation in recent years in the pharmaceutical area.

6   *Acordão do Tribunal da Relação de Lisboa* (Lisbon Court of Appeals Decision) of May 24 1974 in *Boletim do Ministério da Justiça* no 238, p 277. See also *Acordão do Tribunal da Relação de Lisboa* (Lisbon Court of Appeals Decision) of June 26 1974.

7   See Moreira, Pedro Alves – "The role of equivalents and prosecution history in defining the scope of patent protection" in AIPPI Yearbook (2004) – Report Q175, p 162.

8   The Portuguese Supreme Court of Justice can issue *Acordãos de Uniformização de Jurisprudência* (Jurisprudence Uniformity Decisions) which seek to influence lower courts, but they are not binding.

9   Articles 23 and 42 of the PIPC.

application form and a description of an invention need be filed in order for a priority date to be established.[10] The application must then, however, be completed within one year. As a result of a 1959 law, Portuguese applicants are obliged to file European and International (PCT) applications first in Portugal, thus permitting the Ministry of Defence the opportunity to select those inventions that are of interest to the national defence.

On average, a national patent application takes around two and a half to three and a half years to issue. When a European patent application has been requested for the same invention, it is customary for the Portuguese Patent Office to question the applicant as to whether it wishes to proceed with the Portuguese application or abandon it. Article 88 of the PIPC forbids the double protection of an invention as a national and European patent, and the former will lapse as soon as the latter is validated in Portugal. There is no question that the overwhelming majority of patents valid in Portugal are European patents. Portugal also accepts/grants petty patents (*modelos de utilidade*) for small improvements to the state of the art, but for which there is no major inventive step. These petty patents have a shorter maximum life span and a swifter application-to-grant procedure, but in all other matters are considered equivalent to normal patents.

3. The court system

3.1 Basic outline of the Portuguese court system

Portugal has a relatively complex court system, which is divided into two separate and independent structures, namely the judicial courts and the administrative courts. Outside these two structures are other quasi-courts, the most important being the Constitutional Court that deals with constitutional matters. Portuguese administrative courts essentially follow the French model and deal above all with disputes arising out of the exercise of public power and prerogatives (*ius imperii*), such as taxation, building licences and public contracts. Judicial courts are the generalist courts that deal with most private law issues encompassing crime, employment, family and commercial disputes. Both the administrative and civil judicial courts are organised into a hierarchical three-tier pyramid structure, albeit the administrative structure has far fewer courts in the middle and lower tiers. In the judicial court structure, the bottom level is currently made up of 231 first-instance county courts, each of which is often divided into further generic or specialised sections and divisions.[11] Above these county courts there are five district courts[12] (second-instance or appeal courts) and at the pinnacle of this structure stands the Supreme Court of Justice. Other than in exceptional circumstances, all cases must be tried first in a lower or first-instance court before being heard in a higher-instance court.

---

10   The recent reform of the 2003 PIPC has further implemented the concept of a provisional patent application along the lines of the US model.

11   Government plans are currently in motion to change the organisation of the Portuguese judicial court structure, and it is likely that the present situation will not be in place for much longer.

12   Technically, there are six district courts in Portugal: Lisbon, Porto, Coimbra, Évora, Guimarães and Faro. However, although foreseen in a 1999 legal diploma, this last court (Faro) has never been implemented and the Évora district court continues to have jurisdiction over the Algarve region.

3.2 Courts with jurisdiction over patent-related cases

Commercial courts were reinstated in Portugal in 1999 after an absence of more than seven decades. They are specialist judicial courts that handle a limited number of areas which the law has specifically assigned to them. One of those areas is industrial property rights (ie those foreseen in the PIPC, which does not include copyright).[13] The judges of these courts are experienced in the legal aspects of commercial and intellectual property law, but few (if any) have any sort of expert scientific or technological knowledge.[14] There are currently only two commercial courts in Portugal, one seated in Lisbon and the other in Vila Nova de Gaia, on the south bank of Porto. In accordance with Law Decree 186-A/99 of May 31 1999, both of these commercial courts have jurisdiction over the metropolitan areas around the two major Portuguese cities, where the majority of the country's population and companies are based.[15] However, should civil procedural rules determine that the relevant territorial area is not one of those covered by one of the two commercial courts, then typically it will be the local county court that has jurisdiction.[16] Although commercial court judges have a deep understanding of intellectual property, they suffer the disadvantage of being extremely overstretched and currently have large backlogs. Given this situation, there is some discussion today in Portugal about creating courts specifically for intellectual property rights by 2010.

The Lisbon Commercial Court (*Tribunal de Comércio de Lisboa*) is always competent to hear appeals against decisions taken by the Portuguese PTO.[17] Statistically, these appeals are nearly all related to the grant and refusal of national trademarks, and very few actually deal with patents.

Patent infringement is a crime in Portugal, but one that requires that the plaintiff file a formal complaint for the criminal proceedings to begin. If the Public Prosecutor decides to press criminal charges against the infringer, it is necessary to observe the criminal procedural rules in order to determine which court will have jurisdiction. The general rule regarding jurisdiction in criminal cases is that the court of the area where the crime took place will be competent to judge the case.

3.3 Extrajudicial dispute resolution system

Successive Portuguese governments have, in the last decade, introduced a plethora of extrajudicial dispute-resolution systems, such as voluntary[18] and obligatory

---

13   Article 89, para 1 line (f) of Law no 3/99 of January 13 states specifically that commercial courts will deal with all declaratory cases that directly involve industrial property rights in any form foreseen in the Industrial Property Code. Line (h) of the same article also establishes that the commercial courts will declare industrial property rights void or annulled, as foreseen in the aforementioned Code.

14   A patent case heard at a first-instance court will, as a general rule, be judged by a single judge. Higher-instance courts will be composed of at least three judges and in some cases more.

15   The Lisbon Commercial Court's jurisdiction covers the counties of Almada, Amadora, Barreiro, Cascais, Lisboa, Loures, Mafra, Moita, Montijo, Oeiras, Palmela, Seixal, Sesimbra, Setúbal, Sintra and Vila Franca de Xira; while the Vila Nova de Gaia Commercial Court covers Espinho, Gondomar, Maia, Matosinhos, Porto, Póvoa de Varzim, Valongo, Vila do Conde and Vila Nova de Gaia.

16   *"The general rule is that the court with territorial competence is the Civil Court of the district where the defendant has its head office. However, if the defendant is a foreign company the action must be filed at the Civil Court where the plaintiff has its head office, and if both the plaintiff and the defendant are foreign the competent court is the Civil Court of Lisbon."* CRUZ, Nuno – "The role and function of experts in patent disputes" in AIPPI Year Book (1998/111) Report Q136 p 156.

17   Articles 39 and 40 of the PIPC.

18   Currently regulated by Law no 31/86 of August 29 1986.

arbitration courts and magistrates' courts (*Julgados de Paz*).[19] The decisions issued from these arbitration courts are binding. However, one can still appeal to the judicial court system that runs parallel to it. Provided both parties show good faith and are willing to submit themselves to the authority of the arbitration, it is the ideal way to resolve disputes in a swifter and simpler manner. It is worth noting that the 2003 PIPC specifically introduced an ad-hoc arbitration system to deal with the resolution of disputes with the Portuguese PTO, but after five years it was evident that this had not been as successful as initially hoped.[20] There were particular difficulties, such as obtaining the acceptance of the other interested party in the decision (should there be one), and a written authorisation from the Minister of Economy binding the Portuguese PTO. These problems have recently been overcome with the 2008 reform of the PIPC and the creation in October 2008 of Portugal's first permanent voluntary industrial property arbitration centre and the commitment of the Portuguese PTO automatically to accept the rulings of this centre.

The PIPC section on patents has a curious exception to the general rules over jurisdiction in patent litigation cases. Article 59 paragraph 6 of the PIPC makes it obligatory for the parties to resort to arbitration in disputes in relation only to compensation for the contribution given towards the development of an invention. This is actually an area where there is a great deal óf litigation in Portugal. Many novel products or methods are often developed by companies at the request of others without a formal contract specifying patent rights, or additional compensation should the product achieve major market success.

Arbitration is undoubtedly swifter than judicial court proceedings and the decisions are, from a legal perspective, just as reliable. On the downside, deliberate patent infringers will typically not consent to arbitration, since they know that the current delays in Portuguese courts will benefit them.

4. Procedure and timescale of proceedings

4.1 When is it necessary to litigate in civil courts?

Experience has shown that the overwhelming majority of patent infringement cases in Portugal result from a lack of knowledge of the existence of patent protection and the extent of rights granted. An initial warning letter is therefore highly recommended, given that it is often all that is necessary to obtain the infringer's compliance or obtain evidence that clearly indicates that there is no real infringement at all, thus avoiding lengthy and costly litigation. However, when planning to proceed with a warning letter it is always a good idea to obtain as much evidence as possible before contacting or alerting the infringer. Taking simple precautions such as having a public notary certify that on a certain date a website had certain content can sometimes make the difference between a future conviction or acquittal.

There are basically three main types of patent-related declaratory actions that can be filed with Portuguese civil courts, namely:

---

19   Regulated by Law 78/2001 of July 13 2001.
20   Articles 48 to 50 of the PIPC.

- appeal against Portuguese PTO decisions;
- revocation or annulment actions; and
- patent infringement cases in which the plaintiff requests that the court sentence the infringer to cease his infringing activities or abstain from infringing in the future and, when appropriate, pay a just indemnification.

It is also possible to obtain a declaration of non-infringement from a court of law, but the time taken to obtain such a decision almost always proves economically unviable for the interested party to proceed in that way. All these court actions essentially follow the same procedure and the outline below seeks to give a very broad insight into the stages of such proceedings. Again, it must be stressed that despite recent governmental attempts to simplify procedures, Portuguese civil procedural law continues to be highly complex, and it is important to obtain legal counsel with experience and knowledge in these matters.

4.2   Initiating proceedings

(a)   *Who can be a party to the proceedings?*
Formerly, in patent infringement litigation, it was considered that only the patent owner had legitimacy and interest in initiating proceedings and therefore it was essential to obtain the patent owner's intervention. Since April 2008, with the implementation of the Enforcement Directive in national law, it is now possible for licensees and other interested parties to initiate patent infringement proceedings. Revocation or annulment actions can be filed by any interested party and appeals against Portuguese PTO decisions can be filed by those involved in the administrative opposition proceedings, or by any other natural or corporate entity that proves it is directly affected by the PTO decision. As mentioned above, in the latter case, the Lisbon Commercial Court is the proper and competent venue to hear such cases. There has been some debate about whether the Portuguese PTO is or can be a party to these proceedings, but in any case it is always notified of the filing of the action and is obliged to contribute its opinion and forward the relevant file to the court.

(b)   *Deadlines for filing actions*
Infringement and revocation actions have no established deadlines, although it is an obvious precondition that the patent which is being defended or attacked is in force. Appeals (*recursos contenciosos*) against Portuguese PTO decisions must be filed within two months of the publication of the decision in the Portuguese *IP Bulletin* or of the issuance of the final decision by the PTO (should a review of the first decision have been requested).

(c)   *Initial claims and counterarguments*
Civil proceedings will open with the filing of a *petição inicial* (initial petition) or a *recurso contencioso* (appeal), which consists of a document in which the plaintiff will state his version of the facts, the legal rules that are considered applicable and the

claims upon which the court is asked to decide. As will be discussed below, all documented evidence should be presented at this stage. The defendant will then have 30 days as of the date in which he is notified to reply and can do so either by refuting the plaintiff's arguments or by presenting some exception as to why the case should not proceed. The plaintiff can only respond (*replicar*) at this stage if the defendant's defence is based on some exception. This should be done within 15 days of being notified of the reply. The defendant will have an additional opportunity to respond (*treplicar*) if there is a change in the claims made by the plaintiff or he invokes some exception in a crossed action that the defendant may have filed in the meantime. This initial stage of the proceedings is called *articulados* (group of articles), because the petitions filed by the parties must be structured into numbered articles. Once this phase is over, the case will be delivered to a judge so that he may begin to intervene in the proceedings.

4.3 Summarisation and improvement

Once the *articulado* stage has finished and both parties have provided their written versions of the facts, the case is put before a judge whose first task will be to eliminate all issues that are irrelevant to the case. The judge will also resolve any procedural or jurisdictional questions that have been raised. There is no set time limit for the judge to issue his decision on these matters. However, after the summarisation and improvement of the case has ended, the judge will schedule a preliminary hearing within 30 days with the objective of reaching a mutual understanding and, if that is not possible, to decide on things such as the scope of the issues and the evidence that will be discussed in trial. The judge may choose to skip this preliminary hearing phase if he feels it is unnecessary. This stage of the proceedings ends with the issuance of an 'improvement decision' (*despacho saneador*).

4.4 Evidence

The evidence presentation stage (*instrução*) formally begins in the middle of the proceedings, but in reality evidence is provided continuously from the beginning. One fundamental characteristic of Portuguese civil procedural law that must be kept in mind is that the facts presented by the plaintiff must be refuted by the defendant if they are untrue. Should these facts not be expressly or implicitly rejected by the defendant, there is a considerable risk that those facts will automatically be considered proven. There are a few important exceptions to this rule, such as those facts which the law requires be proven by documentary evidence. Nevertheless, the rule of thumb is that the defendant should always contest.

(a) Documentary evidence

As a basic overall rule, documentary evidence must be presented at the *articulado* stage of the proceedings.[21] Additional documentation can be presented until the end of the discussion of the facts during trial, but the party may be liable to pay a fine

---

21 Documentary evidence is to be interpreted in the widest sense possible, given that it includes not only paper documents, but all other physical evidence such as photographs, recordings and objects.

unless it can prove that it was not possible to present the documents at the earlier stage. It is possible to obtain documents held by a third party, although it is necessary to provide convincing evidence or arguments that those documents exist. The recently implemented Enforcement Directive has now greatly reinforced the means by which patent holders can obtain documentary evidence from infringing parties. Should the document not be delivered, there are several means which a patent holder may adopt to force compliance. However, 'automatic disclosure', does not exist in Portugal. Each party is responsible for presenting the facts and documents that support their case.

Written expert opinions can be presented at any phase of the proceedings, except when they are clearly untimely.[22] In Portuguese patent litigation cases, *pareceres* (written opinions) from university professors or other experts on aspects regarding the state of the art, or the equivalence of technical features in two opposing apparatus or methods, are frequently presented. It must be stressed that Portuguese case law states that proof of ownership of a patent must be made by presenting the original letters patent document issued by the Portuguese Patent Office.[23] Relevant foreign documents should be translated into Portuguese and, if they are official documents, it is wise to have them legalised and authenticated by the competent consular bodies. Electronic documents with some sort of digital certification are considered equivalent to signed private documents.

(b)   *Experts*

Expert opinions are also often required and crucial for the outcome of patent cases. However, they are not binding on the judges, and the court has total freedom to interpret and assess the expert findings as it sees fit. Experts are generally appointed by the court to provide technical assistance (eg patent examiners of the Portuguese PTO or engineers of official state laboratories), although the parties can have some input in suggesting and deciding who those experts will be.[24] The parties or the judge can request that a team of up to three experts be appointed. Normally, one is appointed by each party and a third by the court. When an expert opinion is requested by one of the parties, it is necessary to indicate precisely the scope of the matter which will be subject to expert analysis; otherwise the request will be dismissed. The judge can refuse to hear experts requested by the parties if he feels it is unnecessary, or he believes the request has been made merely to delay proceedings. The conclusions of the experts appointed by the court are written down in a report. The parties and the judge have an opportunity to object, or request that the expert report be perfected in some way. Cross-examination of experts is not foreseen in Portuguese civil procedural law, but a second expert opinion may be requested within ten days of being notified of the conclusions of the first opinion.

---

22   Article 525 of the Portuguese Civil Procedural Code. As regards judicial proceedings at the appeal stage (at either the District Court or Supreme Court of Justice), article 706 para 2 states that these opinions must be filed by the time the case is put to the judge for the decision to be issued.

23   (*Acordão do Tribunal da Relação do Porto*) Porto Second Instance Court Ruling no RP199011070123918 of November 7 1990.

24   The parties can oppose the appointment of particular experts by the court if they can show that there will be some sort of bias or that the expert is unsuitable.

(c) Witness evidence

Witness evidence is relatively rare in patent litigation in Portugal, since most evidence can be presented through documents and typically through expert analysis, police reports and so on. The rules regarding witness evidence are generally the same as those in other countries although in civil actions there is a limit to the number of witnesses that can be presented and there are other situations where witness evidence can be denied. Cross-examination is possible, as is hearsay. According to the code of conduct rules of the Portuguese Bar Association, lawyers must not come into contact with witnesses before trial and coaching is prohibited. Witness evidence is more relevant in the relatively few cases where it is necessary to show lack of novelty (eg proving that a certain invention was on show at a certain trade fair), or in disputes over inventorship.

(d) Confessions and other forms of evidence

In addition to those forms described above, other types of evidence are accepted in Portuguese civil courts. These include confessions, depositions, court inspections and legal presumptions which shift the onus of proof from one party to the other. The legal presumptions that are central to patent revocation and infringement actions are the presumption that a granted patent meets the patentability criteria and that novel products that result from patented processes are presumed to have been made by that process unless proved otherwise. It should also be noted that foreign court decisions on the same matter (precedence) are useful as an influence on the Portuguese court, but they carry no official value.

4.5 Final hearing and court ruling

It is during the final hearing that the evidence is discussed and both parties present their closing statements (either written or oral). The proceedings end with the issuance of the court's ruling. If the judgment is not given verbally by the judge at the end of the trial, it should be issued within 30 days but, as in other areas, this deadline must be interpreted as indicative and not binding on the judge. The judgment must be structured in a specific way and the grounds for the decision must be clearly disclosed, indicating what evidence was taken into consideration. The court ruling will also decide on court fees, namely which of the parties should bear them and if they are to be divided amongst the parties (ie what proportions will be borne by the each of the parties). The judgment will be recorded in the court's archives and may serve as the basis for an enforcement action, should the losing party not comply.

4.6 Average length of proceedings

It is no secret, domestically or abroad, that with just a few notable exceptions, Portuguese courts are slow in comparison to many other European counterparts. Although no official statistics have been presented on this matter, experts consider the average time for an intellectual property case to be resolved in a Portuguese judicial court to be around four to five years.[25] The Lisbon Commercial Court, which

---

25   "Um país que não respeita os criadores não merece estar vivo: Entrevista a Manuel Lopes Rocha" – *Jornal de Negócios*, May 2 2008, p 14.

hears the vast majority of patent and trademark-related cases, is overstretched and has a large backlog. The Portuguese state has been criticised several times by international watchdogs and the European Court on Human Rights regarding the delay in obtaining justice and, despite many reforms over the past few years (ie the emphasis placed on implementing paperless files and reducing red tape), the truth is that the effects are not yet visible to the average citizen.

5.   Appeals

Appealing first-instance court decisions is not an absolute right. Portuguese civil procedural law establishes a fairly wide range of restrictions regarding appeals, namely limiting this right to a 30-day deadline (as a general rule) and only for cases over a certain value, and in which there has been a substantial difference between that which was claimed and that which was awarded by the court. With only a few exceptions, appeals to the Portuguese Supreme Court of Justice are restricted to questions of law. It is possible to 'leapfrog' a second-instance court directly to the Supreme Court when a set of specific conditions are met and where that which is appealed is strictly a matter of law. The general rule is that appeals do not affect enforcement. However, one can request the suspension of enforcement when it is argued that enforcement will cause the losing party grave harm and some collateral is given until the decision of the appeal is known.

According to the Portuguese Civil Procedural Code, court actions that deal with intangible interests are considered to have a value that enables them to be appealed to the District or Second Instance Court. Since patents are intangible assets, an appeal to a Second Instance Court will always be possible and a further appeal to the Supreme Court of Justice is possible, should there be some matter of law that needs to be clarified.

6.   Availability of interim relief (especially injunctions)

6.1   Injunctions

Injunctions (*providências cautelares*) and other interim relief measures are available in Portugal. However, until very recently (April 2008), these interim relief measures followed the normal non-specific procedural rules established under the Civil Procedure Code and the PIPC. As previously mentioned, the Portuguese Parliament recently passed legislation implementing EU Directive no 2004/48/CE of April 29 2004 on the enforcement of intellectual property rights, thus bringing about the third alteration of the PIPC since it was approved in 2003. As a result of the implementation of this Directive, Portuguese IP enforcement procedures have been substantially reinforced and harmonised with those of its European partners.

Following the basic principles of interim relief that are applied in other European jurisdictions, recourse to injunctions is accepted whenever it is shown that there is a likelihood of the existence of a right (*fumus boni juris*) and an imminent threat that cannot easily be remedied if there were a considerable delay in the adoption of measures to prevent the harm caused by the infringement (*periculum in mora*). Injunctions are accepted with a summarised account of the claims and available proof, and can be filed before, or at any time during, the main proceedings. The injunction can be authorised by a judge without previously notifying or hearing the defendant when there are convincing arguments or proof that a failure to grant may lead to the destruction of vital evidence or unrepairable damages. If the injunction is filed before the main definitive action, it is necessary to file the main action within 30 days, otherwise the injunction will lapse. Portuguese procedural law does not contemplate the existence of 'permitted delay' and an interim injunction can be filed at any time, provided the above-mentioned conditions are met. Equally, there is no provision in the law for 'protective briefs', but for matters such as patent disputes it is normal for the court to call an *inter partes* hearing before any decision.

6.2 Seizure of assets

Another main type of interim relief foreseen in the PIPC and civil procedural law is the seizure of assets (know as an *arresto*). The purpose of the *arresto* is twofold:

- to provide some sort of guarantee[26] for the payment of damages should the final court ruling be in favour of the plaintiff; and
- to seize the infringing goods or instruments and machines that can only be used for producing the infringing products.

In general, the *arresto* is authorised without prior notice to the infringer, provided sufficient evidence is given of the infringement and of the existence of valid patent rights.

6.3 Seizure of evidence

Although linked to the section on evidence and not clearly identified as an interim relief measure, it is worth pointing out that the PIPC now specifically foresees the possibility of requesting urgent seizure of evidence. This can include a detailed description of the products made available at a certain location, or the collection of samples and physical evidence to present in court, later. Whenever appropriate, this procedure can also be used to seize the materials, instruments and documents used in the infringement of IP rights.

6.4 Border control measures

Portuguese customs police may, on their own initiative, seize or withhold all products that they believe infringe IP rights. The seizure may also be at the request of interested parties such as patent holders. In order for this procedure to be effective it is essential to provide the customs police with the necessary information[27] and, when notified of the seizure, file the necessary court proceedings within ten days.

Despite injunctions and asset seizures being considered urgent procedures which

---

26   The assets seized for the purposes of providing a guarantee for eventual compensation can cover any moveable or immovable asset and include bank account deposits. To discover the assets owned by the infringer, it is necessary to obtain the relevant information from the registry offices. Judicial authorisation is normally necessary for the purposes of obtaining bank information.

27   The customs control procedure is regulated by Law Decree 360/2007 of November 2 2007. The information which must be delivered to the customs police includes special forms and certified copies proving ownership.

should be decreed within a short time limit, the fact is that these deadlines are not always respected by the courts, thus frequently leaving the plaintiff in a very frustrating position. If a party is certain that infringement is taking place and wishes to take immediate action, a criminal complaint is sometimes the fastest way to obtain results. To ensure maximum speed, the complaint should be made directly to the public prosecution representative of the local court.

7.   Available remedies

To date, when patent infringement has been proven to have existed in a court of law, obtaining a favourable judgment from the courts has been relatively straightforward. Under Portuguese civil procedural law, the court can only sentence the defendant to that which has been requested by the plaintiff. A judge can never sanction the losing party for something which has not been claimed, nor to higher financial compensation than that which the plaintiff has requested (although the judge can decide on a lesser amount and is often asked to decide based merely on reasons of equity).

Typical remedies consist of ordering the infringer to cease his infringing actions, an injunction prohibiting the defendant from infringing those patent rights in future, and also the delivery up and/or destruction of any infringing products or instruments. When it comes to financial remedies, experience has shown that Portuguese courts have been rather reluctant to order the defendant to pay damages, or in any case, monetary reparations tend to be symbolic.

The Portuguese Civil Code stipulates that indemnities are the means by which a party who has suffered damage is restored to the situation he would have been in had the illegal action never occurred. Financial remedies are calculated by taking into consideration direct economic losses, lost income and moral damages Punitive or exemplary damages are not accepted in Portugal and therefore only that loss which can actually be proved to have been suffered or is considered to be lost revenue will be taken into consideration. This is the first problem for most patent infringement cases, given that plaintiffs have few (if any) viable ways of proving that the infringer's actions actually affected sales or were detrimental to the plaintiff's image. Not only is there difficulty in proving damages and lost profits but, more crucially, it is often impossible to show the causal link between the infringer's activities and the losses claimed. Most Portuguese judges have not made matters easier, as they often adopt a strict view in terms of proof and reasonable claims. It is also commonplace to find that the plaintiff does not have adequate records of his sales and profits, or even a clear idea of the economic impact of the infringement on his business, and this obviously makes it extremely difficult argue persuasively in court. In addition, Portuguese courts have a tendency to disregard claims for damages based on moral and image rights when it comes to patent infringement, making reparation particularly challenging.

However, all this is set to change. The new rules brought about by the Enforcement Directive, although still very fresh and untested, will assuredly cause a small revolution in terms of patent enforcement and compensation for infringement. Certainly, one of the more important measures recently implemented is the calculation of compensation based on the theoretical rate of 'reasonable royalties' should it be impossible to prove damages, as well as the inclusion of the costs of prosecuting the case. In one fell swoop, the Enforcement Directive has cured the main affliction of Portuguese patent litigation over recent decades, namely the lack of compensation when financial damages cannot be proved in a court of law.

Parallel to the civil remedies, criminal law makes patent infringement punishable by up to three years in prison, or a fine of 360 days.[28] The court will almost invariably apply a fine rather than a jail sentence for IP infringement. As a 'semi-public' crime, patent infringement is investigated and prosecuted by the state, through the *Ministerio Público* (public prosecution). It is necessary, however, for the plaintiff to file a formal complaint in order to initiate the proceedings. The complaint can be filed either with the police or the public prosecution representative of the nearest court. Portuguese criminal procedural law permits the plaintiff to request financial indemnification for the infringement of his rights in the same proceedings that will decide on the criminal charges. Should the plaintiff wish, it is possible to file a separate action in the civil courts that will run parallel to the criminal case.

8.   Costs

As in most countries, patent litigation in Portugal is fairly expensive. Given that it is a small niche practice area in Portugal, relatively few lawyers and law firms can genuinely regard themselves as expert in IP law, and even fewer can truly claim to grasp patent-related issues. The cost of retaining a lawyer or litigation team specialising in patent law is consequently much higher than the going rate for a generic non-specialist lawyer. If one adds fees for technical expertise and takes into consideration the large number of hours necessary to analyse a patent case and defend it in court, it is easy to see why patent litigation tends to be more expensive than other types of litigation (eg employment issues). The cost of patent litigation in Portugal varies tremendously from case to case due to the complexity, time, size of the litigation team and so on. Consequently, it is not possible to indicate an average. Court fees for prosecuting a patent-related case will also fluctuate depending on the complexity of the proceedings and how they develop. Nevertheless, it is possible to state that litigating in Portugal is generally far less expensive than other countries such as the United States or the United Kingdom, where hourly rates tend to be far higher than those in Portugal.

As a general rule, the losing party pays the winner's court fees. In those cases where the declaratory action does not have a loser, it will be the person who benefited from the action that will bear the court costs. If there are various losing parties, costs will be divided among them. Traditionally, the losing party was also responsible for the payment of legal fees, but these were decided by the judge based on an official statutory scale rather than actual expenditure. The difficulty in determining the actual damages suffered meant that, in most cases, the plaintiff

---

28   Each day for the purposes of fines can vary between €1 and €498.80 depending on the court's decision on factors such as the economic resources of the person or company convicted of the crime. The maximum fine applicable by the state is therefore €179,568. This does not include the amount due in damages, which is determined separately.

would request that the court decide compensation based on values of justice (so awards tended to be very small). When an attempt was made to include legal fees in the assessment of damages, courts in the past often considered that they did not fit the interpretation of indemnification. Recently introduced article 338 L of the PIPC now implicitly foresees the inclusion of legal fees in the compensation request, and in future it is expected that patent litigation in Portugal will be far more worthwhile (from a cost perspective). However, there continue to be some doubts as to whether recoverable legal costs will actually match expenditure, and it will only be possible to answer this question in years to come.

9.   Hot topics

It is an undeniable fact that Portugal has a very low level of patent litigation, particularly when compared with other larger European partners. Recently, there has been a surge in pharmaceutical patent litigation, but it has been directed mainly at the Portuguese authorities that grant market authorisations for generic drugs while patents are still valid (the Bolar exception).

The lack of case law in Portugal can be argued to be advantageous in most cases. The fact that there are no real established interpretations or tests applied by Portuguese judges, which admittedly leads to some uncertainty, also permits a great deal of leverage as to how one can argue the case and what foreign case law should be included as examples to follow. In other words, patent litigation in Portugal allows for a great degree of freedom in deciding what interpretations and decisions would best apply to a particular case.

The recent implementation of the Enforcement Directive and the 2008 reform of the PIPC has raised high hopes and expectations among IP experts in Portugal. For decades, companies and individuals had been somewhat reluctant to protect their IP rights in Portugal, because there was a feeling that it was not worth the trouble, given that enforcing those rights was often an unrewarding experience. With the new laws now in place and an extremely active enforcement police (ASAE), the tide is finally turning and the perception is that infringement will decline drastically. Lawyers have finally been given effective procedural measures which they can apply, placing Portugal on a par with the rest of its main European counterparts.

The Achilles heel continues to be the long delays in prosecuting cases in court. Successive Portuguese governments have attempted to tackle this problem, and the recent creation of a permanent voluntary IP arbitration court is a welcome move in that direction. In conclusion, patent litigation in Portugal today offers essentially the same rights, guarantees and procedures as those in countries that have a stronger patent litigating culture and experience. One needs just a little more patience in Portugal.

Spain

José Massaguer
Ingrid Pi
Uría Menéndez

Patent litigation in Spain is governed mainly by the Spanish Patents Act[1] and the Spanish Act on Civil Proceedings.[2]

Whilst the Spanish Patents Act deals for the most part with substantive issues relating to patent litigation, the Spanish Act on Civil Proceedings governs most of the procedural issues. Nevertheless, there are a few highly important procedural aspects in patent litigation (eg jurisdiction, venue and licensees' standing) which are governed by the Spanish Patents Act.

Both acts were amended in order to implement the provisions of the IP (intellectual property) Enforcement Directive.[3] As a result, Spanish law has an extremely efficient and comprehensive set of proceedings for patent litigation, as well as an effective set of remedies.

Since 1986, Spain has been a member of the European Patent Convention. Since then, there has been a *de facto* harmonisation between Spanish law and the European regime with regard to the main issues relating to patentability requirements and prohibitions, as well as the Convention on the Community Patent with regard to issues relating to the protection granted to a patent. The Spanish Patents Act was amended in 2002 to implement the EU Directive on the protection of biotechnology inventions and in 2006 to include the 'Bolar exception' as a particular instance of the exception for experimental use.

1. The court system

1.1 Spanish courts for patent litigation
Patent litigation matters are dealt with by specialised courts in the first instance and, in some cases, in appeals (eg Madrid, Barcelona and Valencia). Specialised patent courts are the so-called commercial courts, which also deal with, among others things, unfair competition and copyright matters.

Experience with these specialised courts, which were created in 2004, has so far been fairly positive, particularly in terms of speed, a sensitive approach to the nature of the evidence, the assessment of the different aspects involved in patent cases (in particular with regard to litigation on pharmaceutical patents), and the legal quality of the decisions.

---

1   Law 11/1986, dated March 20 1986 on Patents.
2   Law 1/2000, dated January 1 2000 on Civil Proceedings.
3   Law 19/2006, dated June 6 and the Directive 2004/48/EC of the European Parliament and of the Council of April 29 2004 on the enforcement of intellectual property rights.

Although Spanish judges have no technical education, both technical and legal issues involved in patent litigation are in fact carefully addressed and analysed by the courts. Indeed, judges carefully review the expert evidence produced in the proceedings in order to give themselves an overall picture of the technical questions underlying the case. Accordingly, court experts have become a key element in patent litigation in Spain.

There are 17 courts which deal with patent matters in Spain, both at first instance and in appeals. These are the commercial courts located in the city where the High Court of the Autonomous Region of the territories is based. Nevertheless, most of the patent lawsuits are handled by the Barcelona or Madrid commercial courts. The specific commercial court jurisdiction over particular proceedings depends, on the one hand, on the type of proceedings (eg infringement, revocation or description proceedings) and, on the other hand, on the defendant's domicile and/or the place where the patent infringement was committed or where its effects were, or are, taking place. This is examined below.

1.2 Spanish litigation teams

In Spain, lawyers have a right of audience in courts, provided that they are members of a Bar Association. In contrast to other jurisdictions, such as the United Kingdom, the Spanish system allows all lawyers rights of audience.

However, a peculiarity of the Spanish system is that in order to appear before the court it is requested that the parties are represented by a court agent. The court agent is in charge of representing the party before the court and liaises between the court and the party, is notified of the court decisions, files writs to the court and is served writs submitted by the counterparty.

Due to the fact that access to the legal profession is via law school, Spanish lawyers do not have a scientific education. However, there are well-known patent litigation specialists in Spain. In order to prepare cases, the technical advice of scientists is a key issue. Accordingly, patent litigation teams tend to include experts who, as stated above, not only help the patent team to prepare the case but also explain technical issues to the judge.

In contrast to other jurisdictions, Spanish lawyers are responsible for both the factual and legal arguments of the case and there is no patent attorney in charge of the factual technical issues. However, patent attorneys may act as independent experts in the proceedings, or may also be engaged as independent consultants to advise the legal team and to prepare expert reports on the technical issues involved in the litigation.

2. Patent actions

2.1 Patent holder's actions

(a) Patent infringement

From the patent holder's perspective, it is important to analyse which acts constitute a patent infringement, as well as the actions available to the patent holder in case of infringement.

The Spanish Patents Act entitles the patentee to prohibit non-authorised third parties from acts of direct and indirect infringement. Direct patent infringement consists of manufacturing, offering for sale, commercialising or using the patented product, as well as importing or possessing the product for one of the above purposes. In turn, with regard to process patents, direct patent infringement consists of using the patented process or offering its use, when the third party is aware, or the circumstances make it obvious, that use of the process without the patent holder's consent is prohibited. In addition, offering for sale, putting on the market or using the product directly obtained through the patented process, or importing or possessing it for any of the above purposes, also constitutes a direct patent infringement.

Indirect patent infringement consists of handing over, or offering to hand over, to unauthorised persons any elements relating to an essential part of the invention to be used for working the invention, when the third party is aware, or the circumstances make it obvious, that such elements are appropriate to work the invention and are to be used for that purpose.

Under Spanish law, the patent holder is entitled to file infringement actions in the event that the patent infringement has already been committed, and also if the infringement practices have been prepared and the actual infringement is imminent.

(b)   ***Patent infringement: reversal of the burden of proof***
For process patents it can be difficult to obtain appropriate evidence on infringement (ie in certain cases it is difficult and/or time consuming to have access to the process used by the defendant). Therefore, in order to facilitate the protection of the interests of certain process patents, the Spanish Patents Act sets out the presumption that, if the subject matter of a patent is a process for obtaining a new product or substance, any product or substance possessing identical characteristics will be considered to have been obtained by using the patented process in the absence of evidence to the contrary. Accordingly, this presumption will only occur if the claimant duly evidences:

- the existence and enforceability of the patent;
- subject matter which is a process for obtaining a new product or substance; and
- the presence of either a product or a substance possessing identical characteristics to substances or products directly obtained by carrying out the patented process.

The defendant can rebut the presumption either:

- by proving that the product directly obtained through the patented process lacks novelty, or that a potential infringing product does not have the characteristics of the product obtained by carrying out the patented process; or
- by providing evidence that the process used differs from the process patented by the claimant.

The standard of proof requested to rebut the presumption differs from main proceedings and pre-trial relief proceedings.

With regard to main proceedings, courts have been asked to issue a decision regarding the kind of evidence that the defendant must provide to avoid the presumption and, in particular, on whether the defendant is obliged to evidence the process actually used to manufacture the product in question.

The findings of the courts in relation to this issue have been made on a case-by-case basis and no general doctrine has been provided by case law. However, in most of the cases in which the defendant is considered to have evidenced that the process actually used differs from the patented process, the defendant has produced evidence on the process actually used (eg certificates issued by independent parties after reviewing the premises where the allegedly infringing process was used, or a copy of the public part of the Drug Master File filed with the health authorities describing the synthetic route of the active principle).

Nevertheless, in certain cases the courts have found that the disclosure of an alternative process is sufficient evidence for this purpose, and the defendant has not been requested to produce evidence of the actual use of this process to manufacture the alleged infringing products. A recent case considered that: (i) a notarial affidavit from the manufacturer of the active principle stating the process used; and (ii) a report issued by the University of Barcelona stating that the compound in question could be obtained through over 22 processes, is evidence that the product imported was obtained through an alternative process other than that which was patented.

In interim injunction proceedings, courts have stated that the presumption can be rebutted if, as a result of the evidence provided by the defendant, it is questionable that the process used by the defendant is infringing the patent (although mere doubt on the infringement is insufficient to support a finding that there is no infringement within the main proceedings). Therefore, the allegation and evidence of the existence of (industrially and commercially viable) alternative proceedings could suffice to rebut the reversal of the burden of proof within the interim injunction proceedings.

(c)　***Remedies on the merits***
The following remedies can be sought for patent infringement:
- an injunction to cease;
- compensation for damages;
- seizure of infringing products, devices and means chiefly used to carry out the infringement;
- transfer of the ownership of the objects and means which have been seized (in which case the value of the relevant products will be deducted from the total amount of compensation for damages);
- adoption of the necessary measures to prevent continued infringement of the patent and destruction of the infringing products and means previously seized; and
- publication of the court's judgment.

In accordance with the provisions of the IP Enforcement Directive, the patentee is also entitled to institute actions against third parties who, although not carrying out the patent infringement themselves, provide services to the infringing party which are being used to commit the patent infringement. In particular, patent holders can request that the court orders these third parties to cease providing the services, or to destroy any infringing products they may have. This provision does not affect the specific regime provided in e-commerce regulation in relation to providers of information services.

With regard to damages, compensation includes the actual harm suffered by the patent holder in addition to lost profits, and any costs incurred by the claimant in order to verify that the patent was being infringed. The amount of compensation for loss of profits is determined according to the negative economic consequences of the infringement, or the amount which would accrue under a hypothetical licence, at the choice of the claimant.

The criteria for the negative consequences of the infringement include loss of profits, account of profits and moral damages. However, it is uncertain whether both loss of profits and account of profits can be claimed. In any event, in order to establish compensation, the profits obtained by the infringing party from a product where the essential part (from a commercial point of view) is the patented invention can be taken into account.

In order to fix the amount of the hypothetical licence, the prestige of the infringed patent, its duration, and the licences potentially granted should be taken into consideration.

A peculiarity of the Spanish system is that the claimant's right to choose the compensation criterion is subject to proof that the patent at hand is being exploited in any member country of the World Trade Organisation. In the absence of this evidence, compensation must be determined in accordance with the hypothetical licence criterion.

As a general rule, under Spanish law actions for damages are subject to the ability to prove that the claimant has suffered actual damage as a result of a third party's infringement and the amount of such damage. As an exception to this rule, Spanish courts have maintained that in the event of patent infringement it should be presumed that the patent holder has suffered damage, although its amount must be fully evidenced by the claimant during the proceedings. It is not possible to defer the fixing of damages to separate proceedings unless the claimant evidences that it is unable to fix such amount in the proceedings (and courts tend to be very restrictive in this regard). Furthermore, it is worth noting that Spanish law entitles the patent holder to ask an expert to review the defendant's accounts in order to establish the amount of damages.

Only the importer and manufacturer of the infringing product and the party who uses the infringing process are subject to a strict liability rule. Other potential infringing parties will only be liable for damages if:

- they were sent a cease-and-desist letter by the patent holder informing them of the patent right and requesting them to cease infringing its exclusive rights; or
- the defendants acted negligently (ie if they were aware of the unfairness of their exploitation practices).

(d) Pre-trial relief remedies

Pre-trial relief can be sought against any potential defendant in patent infringement proceedings (ie including the infringing party and the third parties' service providers) in the case of either actual or imminent infringement (eg once preparatory acts have been completed, so that the actual infringement may be carried out at any time).

The Spanish Patents Act provides specific pre-trial relief remedies in the event of patent infringement, in particular:

- cessation of the infringing practices;
- withholding and storing of the infringing products and the means used exclusively in the manufacture or use of the patented process;
- guarantee of any compensation for damages; and
- provisional recording of the filing of the complaint in the relevant registers.

The pre-trial remedies set forth in the Spanish Patents Act do not limit the remedies available to the claimant. In accordance with the Spanish Act on Civil Proceedings, the claimant is entitled to seek any other remedies that are necessary to guarantee the protection of the patent holders in the event that the actions on the merits are granted by the court (eg communication of the ruling granting pre-trial relief to third parties).

2.2 Opposing party's actions

A patent holder is faced with two main patent actions from the infringing parties' point of view: an action for declaration of non-infringement (declaratory action), or a revocation action. The revocation can take the form of either a direct nullity action, or a revocation counterclaim within the infringement proceedings.

(a) Declaratory action

The declaratory action is intended for the purposes of obtaining a court declaration that a particular practice does not constitute an infringement of a specific patent.

This action may be filed by any party holding a legitimate interest. Since the Spanish Patents Act does not define who the interested parties are, this should be assessed and decided on a case-by-case basis. In any event, any third party that could be sued for the infringement of the relevant patent should be considered an interested party in the non-infringing proceedings. Once a patent infringement claim has been filed, the defendants have no standing to file non-infringement actions (since the non-infringement arguments should be alleged and evidenced in the answer to the patent infringement claim).

Only the patent holder may be a defendant in non-infringement proceedings, although licensees will be notified of the filing of the action, so that they may intervene in the proceedings as interested parties.

The action is subject to the fulfilment of strict requirements prior to its filing. In particular, the potential claimant must send the patent holder a notice letter regarding the exploitation practices intended and requesting confirmation that these practices do not constitute a patent infringement. The claimant will only be entitled to file the declaratory action if the patent holder does not respond to the claim within a 30-day term, or if the claimant does not accept the patent holder's answer.

The filing of a non-infringement action and the sending of the notice letter in preparation thereof do not prevent the patentee from filing an infringement action. In particular, the infringement action may be filed either as a direct action, prior to the expiry of the 30-day term, or as an infringement counterclaim, together with the answer to the non-infringement claim.

Finally, the declaratory action can be filed together with a direct revocation action. Should this be the case, the patentee will be granted a 20-working-day term to respond to both the non-infringement claim and the revocation action and, as appropriate, to file the infringement counterclaim. All these actions are to be dealt with in the same proceedings.

(b)   Revocation action

A common strategy for infringing parties to avoid an infringement declaration and related remedies is to file a revocation action with the objective of having the patent declared null.

In most cases, the revocation action is filed as a counterclaim within the infringement proceedings, together with the response writ to the claim for patent infringement. In this case, the patentee (as defendant in the nullity proceedings) is granted a 20-working day term to respond to the revocation action. Both the infringement proceedings and the revocation proceedings are dealt with simultaneously by the same court and a single decision on both infringement and nullity is issued at the end of the proceedings.

However, the potential defendant may try to anticipate the infringement action and file a direct revocation action against the patent holder. This strategy has in practice proved effective in some cases where the nullity argument is intended to be used as the main line of defence within the framework of pre-trial relief for patent infringement. Under normal circumstances, however, the nullity defence in relation to a motion for interim relief is not effective, since the main case law states that the patent should be considered valid within the interim injunctions proceedings.

3.   Procedure and timescale of proceedings

3.1   Preparation of the proceedings: warning letters

Before filing a patent infringement action, the possibility of sending a warning letter must be carefully assessed. The decision on whether or not to send the warning letter should be made on a case-by-case basis.

In general terms, sending a warning letter is advisable in the event that the potential infringing party is carrying out exploitation practices (other than importing or manufacturing the patented product or using the patented process), if the claimant is interested in seeking compensation for damages caused as a result of the infringing practices. In this event, the sending of the warning letter is required to guarantee that these potential infringing parties will be liable for the damages.

On a different note, sending a warning letter is a useful means of interrupting the statutory limit of five years for patent infringement actions. Moreover, the sending of a letter could be advisable to reinforce the fulfilment of the urgency requirement (*periculum in mora*) for pre-trial relief (since some recent precedents have stressed the refusal to accept the undertakings requested in warning letters as indicia of imminent infringement).

Warning letters must be drafted carefully to ensure the fulfilment of their main purposes and to avoid any potential pitfalls. In particular, any content which could lead to the qualification of the warning as an act of unfair competition must be avoided. Indeed, under certain circumstances, statements relating to an infringing third party could amount to a disparaging act (eg reference to the infringing nature of a product made to the retailer).

3.2   The proceedings

For patent litigation, Spanish law provides for two instance proceedings. Only under certain specific circumstances is it possible to appeal against second-instance rulings to the Supreme Court.

First-instance proceedings are mainly based on legal writs and oral hearings, whilst appeal proceedings and Supreme Court proceedings are normally based on legal writs.

(a)   Proceedings before the first-instance court

(i)   Venue

The general rule for establishing the venue is the defendant's domicile. In both declaratory and revocation actions, the claim must be filed before the courts corresponding to the defendant's domicile. In patent infringement cases, in addition to the defendant's domicile, the claim may be brought before the court corresponding to the place where the infringement was committed, or where its effects were (or are) taking place.

(ii)   Standing

Both the patentee and licensees (the latter in limited circumstances) are entitled to institute actions for patent infringement. In particular, registered exclusive licensees have standing to file infringement actions, unless the patentee has excluded this right in the relevant licence agreement.

Registered non-exclusive licensees are not entitled to bring actions independently of the patentee, unless they have been expressly authorised to file actions on their own behalf by the patentee.

Non-registered licensees (either exclusive or non-exclusive) do not themselves have standing to file patent infringement actions. Nevertheless, according to certain case law they are entitled to file these actions (as co-plaintiffs) together with the licensee. Nevertheless, the interests and legitimate interests of non-registered licensees can be protected under unfair competition doctrines.

Furthermore, in the event of infringement, licensees should inform the patentee of the infringement practices in order for that party to take appropriate action. If the patentee refuses to take action, or does not initiate the infringement action within a three-month period following the licensee's notice, the registered licensees (who do not have the standing to institute actions for infringement, according to the rules explained above) are entitled to bring infringement actions on their own behalf. In any event, prior to the expiry of the above-mentioned three-month period, the licensee is entitled to file a motion for pre-trial relief if the infringement activities are causing significant damage to the licensee. The filing of this motion is subject to the sending of the requesting letter referred to above.

(iii)   *Claim motion*

The initial motion of the proceedings is the claim motion. In contrast to other jurisdictions, the claim does not consist of a mere writ declaring that an action for infringement is being brought; but, rather, it must contain all facts and legal arguments supporting the patent infringement action, as well as the relief specifically sought.

The claimant must, therefore, invoke all actions available against the defendant within the initiated proceedings. Where the claimant fails to do this, they will be prevented from bringing these actions from the filing of the answer to the claim (eg if the claimant only seeks an injunction to cease the infringing activities it would be prevented from seeking damages arising from the infringement at a later date. Likewise, if the infringement activities also amount to an unfair practice, the relevant declaration and appropriate remedies must be sought in the infringement proceedings, otherwise the unfair competition action cannot later be instituted). However, the claim can be enlarged after its filing to add new remedies, other facts or other defendants, provided the defendant has not filed the answer to the claim.

As a general rule, the claim must attach all forms of evidence, including expert reports, to demonstrate the facts alleged in the claim. This rule – which also applies to the defendant, who must produce and file the evidence together with the answer to the claim – has very few exceptions, and these exceptions tend to be strictly construed by the courts. In particular, additional documents may be filed under the following circumstances:

- expert reports which were requested before the filing of the initial motion and which could not be provided by the expert on the date of filing of the claim. In order to be accepted by the courts, the party must identify the expert report which it intends to file and must justify the reasons why the expert was unable to provide it by the relevant date. Under this exception, the expert report must be filed one day before the date of the preliminary hearings, at the latest;
- expert reports for the purpose of challenging arguments raised by the defendants in their initial motions (eg in the answer to the claim for infringement, or in the answer to the counterclaim for revocation). These reports must be filed five working days prior to the main hearing;
- documents and expert reports relating to facts which occurred after the initial motion;
- documents which were issued after the filing of the relevant initial motion

(provided that it was not possible to produce them on the date of filing of the initial motion);

- documents which, even if issued before the filing of the initial motion, were acknowledged by the parties after its filing; and
- banking, financial or commercial documents in the control of the infringing party. The claimant may request that the court orders the defendant to produce those documents within the proceedings. In order to be provided with these documents, the claimant must specifically identify the requested documents (this does not amount to a disclosure order), and must provide reasonable evidence that they are in the defendant's possession.

(iv) Granting leave to the claim

Having received the claim, the court analyses it from a formal point of view and, as appropriate, issues a decision granting leave to the claim and summoning the defendant to file the answer to the claim within a 20-working-day period.

(v) Response to the claim/counterclaim

The defendant must file a defence writ within the 20-working-day term. This term cannot be extended unless additional claims have been brought or new facts alleged by the plaintiff.

The defendant will usually file all evidence with the court, together with the response to the claim. The defendant is not entitled to produce further defence arguments and documents at a later stage.

Defendants can allege the nullity of the patent on which the infringement action is based, as a defence. It is uncertain whether or not the nullity defence can be raised in separate, further main proceedings. In any event, the nullity defence in the answer to the claim for patent infringement can be brought either by means of an exemption – the effects of the assessment and ruling on the nullity defence will be limited to the parties to the proceedings – or a counterclaim. (The ruling declaring the nullity of the patent will include an order to cancel its registration and will have effects *erga omnes*.) The nullity exception and the counterclaim for revocation will be dealt with jointly by the court in charge of the infringement proceedings. As a result, the proceedings will not be stayed. The decision on the merits will deal with both the revocation and, if the patent is not revoked, the infringement claim.

(vi) Preliminary hearing

Once the response to the claim (and potentially to the counterclaim) has been filed, the court will issue a summons for the parties to attend a preliminary hearing in an attempt to settle the dispute.

If no settlement is reached, the preliminary hearing deals with procedural matters referring to the claim and the response to the claim which could impede the court making a decision on the merits (eg the joinder of proceedings, legal standing of the parties). In addition, parties are allowed to clarify and expand on their allegations regarding non-essential aspects, to present new facts which have arisen since the filing of the initial motion of the parties and to present related allegations and evidence.

Once procedural issues have been dealt with, the parties, together with the court, define the facts in dispute. Then, on the basis of the facts in dispute, each party proposes the evidence to be produced within the main hearing to the court, and also opposes the forms of evidence proposed by the other party. As explained above, documentary evidence must usually be submitted to the courts together with the claim, or the response to the claim; and within the preliminary hearings the parties only request that the judge takes them into consideration when deciding on the case.

(vii)   *Main hearing*

The purpose of the main hearing is to examine the evidence admitted by the court during the preliminary hearing, including examination and cross-examination of parties, witnesses as to facts and technical experts whose reports should have been previously produced in the proceedings.

Having examined the evidence, an oral pleading is submitted by each party, consisting of final statements and conclusions, on both the results of the evidence produced within the whole proceedings in connection with the disputed facts, and the legal doctrines applicable to the case.

There are no initial pleadings and neither are basic skeleton arguments exchanged between the parties.

(viii)   *Final proceedings to examine evidence*

This is an exceptional procedural step in relation to the acquisition of additional evidence. Within final proceedings, the only evidence that may be produced (at the court's discretion, should it consider that the evidence is necessary to resolve the case) is:

- evidence that could not be produced within the main hearings, provided that the party proposing the extraordinary evidence is not responsible for the failure to produce the evidence within the main hearings; or
- evidence relating to facts arising after the main hearing.

(ix)   *Final decision by the commercial court*

Within a 20-day period following the trial, or the final evidence proceedings, as applicable, the court should, in theory, issue a final judgment. However, due to the workload of the courts, the judgments are normally issued beyond this term.

(b)   **Proceedings before the Court of Appeal**

The final decision issued by the commercial court can be appealed before the Court of Appeal. The filing of the appeal does not stay the enforceability of the first-instance decision and a provisional, compulsory enforcement of a favourable ruling may be sought by the claimant in connection with all the remedies granted except those that are declaratory (eg the revocation of the patent).

The scope of the appeal proceedings is not limited to the review of the legal arguments raised by the first-instance court. The appellate court can review both the facts that have been considered as evidenced by the parties and the law applied by the first-instance judge. The main purpose of the appeal proceedings is to have the first-instance proceedings carefully reviewed in light of the allegations made by the parties, and either to confirm the findings of the first-instance court or partially or totally revoke them.

Consequently, it is not usually possible to file new pieces of evidence. The filing of additional evidence is limited to the forms of evidence which were unlawfully rejected in the first instance, or which could not be adduced in the first instance, or to new facts.

Due to their purpose and scope, the appellate proceedings deal with documentary evidence and, under normal circumstances, no oral hearings are held. However, if new evidence is brought within the appeal proceedings, a hearing will be held in order to examine items of evidence that were admitted by the court, if any, followed by a final, oral pleading of each party. In addition, a hearing may be held at the request of both parties and, in any event, if the appellate court considers it necessary.

(c) **Proceedings before the Supreme Court**

In the case of patent litigation (both revocation and infringement), appeal of the second-instance rulings to the Supreme Court is limited to those cases involving so-called "interest to the Supreme Court". This "interest" arises in cases where there is contradictory case law on the legal principle applied to solve the dispute.

In particular, such interest will arise if the court of appeal has applied a legal principle or a legal rule in contravention of the prior case law of the Supreme Court. Alternatively, it is possible to file an appeal to the Supreme Court if there is contradictory case law of two different appellate courts on the rule that has been applied by the court of appeal to decide the case.

The appeal to the Supreme Court should be announced within the five days following the notification of the judgment of the appeal court to the parties. The appealing party should state that the case involves "interest to the Supreme Court" in this motion and should identify the contradictory case law in its judgment. If the court of appeal considers that requirements for filing the appeal to the Supreme Court are met, it will grant the appealing party 20 working days to file the appeal motion.

Once the appeal motion is received by the Supreme Court, it will issue a decision granting leave to appeal, or rejecting it due to the lack of "interest to the Supreme Court". The Supreme Court will then issue a final decision on the case and this cannot be appealed.

3.3 Potential interference

In Spain, the mere filing and prosecuting of oppositions against European patents before the European Patent Office does not have any effect regarding infringement or nullity proceedings over the Spanish part of the relevant European patents. Infringement proceedings are not stayed and a final ruling is rendered by the relevant court, without awaiting the decision of the European Patent Office on the opposition.

If the opposition decision is issued before a definitive ruling has been enforced, it is possible to invoke the European Patent Office's decision in the relevant proceedings. Therefore, if the patent has been revoked, the defendant could allege this as a new fact and file the relevant evidence within the proceedings (ie first-instance proceedings or appeal proceedings) and the infringement action will be rejected accordingly. If a final decision has already been taken, the defendant could invoke the revocation of the patent within the relevant enforcement proceedings in order to avoid being ordered to comply with the infringement judgment.

If the opposition decision is issued once the final decision has already been enforced, the defendant should initiate appropriate declaratory proceedings seeking recognition of its right to continue exploiting the patented invention and, if the required conditions are met, appropriate remedies in connection with the results of the enforcement of the final decision on infringement.

3.4 Timescale of proceedings

Under ordinary circumstances the timescale of proceedings, from filing of the claim to the main hearings, is approximately 12 to 18 months, including interim measures and revocation counterclaim.

3.5 Costs

In Spain, legal costs (attorneys' fees) are determined in line with statutory scales established by each bar association (although they are currently all quite similar).

The scale of costs depends on the value of the action. In patent litigation, this is usually determined in accordance with the damages sought by the claimant. However, if the value of the action cannot be determined, as is the case for certain patent actions (eg revocation and declaratory actions, or infringement actions if the claimant has not had access to the appropriate data to calculate damages upon filing the action), the scale's maximum is fixed at €18,000. Once a scale has been established, costs are determined by applying a percentage.

In addition to legal costs, the parties are also entitled to recover the amount paid to the court agent. These are also determined in accordance with statutory scales.

Finally, the successful party is entitled to recover amounts paid to experts for preparing relevant reports, subject to certain limitations.

4. Availability of pre-action evidence gathering/disclosure

In Spain, there is no disclosure. Evidence on the infringement has to be obtained, or at least duly identified, before filing the relevant action.

Nevertheless, given that the Spanish system offers patentees very comprehensive proceedings to obtain and/or to secure the evidence on a patent infringement, they are not in a substantially worse position in Spain than in countries in which disclosure exists. Indeed, Spanish law provides patent holders with a set of proceedings that accommodate almost any potential situation in relation to evidence of infringement.

If the patentee needs to verify whether or not an infringement practice is being carried out and/or to obtain the necessary evidence on it, description proceedings can be sought. In contrast, if there is evidence that a patent infringement is being carried out, but the patentee lacks some of the requested information to file the action against the right defendant, or needs further information to define the remedies (eg to decide the indemnification criterion), or to identify potential co-defendants, preliminary proceedings can be initiated. If the patentee has identified evidence of the infringement that is in the defendant's possession, it is possible to compel its production within the proceedings. In addition, if the patent holder is aware of the existence of the evidence and suspects that it could be destroyed during the proceedings (and prior to gaining access to it), proceedings to secure evidence can be started in order to protect it.

It should be noted that not only patentees can benefit from these proceedings. Any party who has standing to file a patent infringement action is entitled to initiate proceedings, including registered exclusive licensees.

The venue for these proceedings corresponds to the court that will handle the infringement proceedings that are being prepared or conducted.

4.1   Proceedings to gather evidence before the lawsuit

(a)   Description proceedings

Spanish law provides patentees with specific proceedings to verify whether their patents are being infringed and to provide them with the adequate evidence to prepare and file the infringement claim. Due to their specific nature (ie to ascertain whether infringement is taking place), these proceedings must be initiated, prosecuted and filed prior to a potential infringement action.

Description proceedings consist of a review of the premises, machines and processes relating to potential acts of infringement, in order to verify whether or not an infringement is being carried out.

Although the specific wording of the Spanish Patents Act limits the scope of description proceedings to the review of premises, machines and processes, they have been extended to obtain samples of potential infringing products, documentary evidence relating to the patent infringement and even witness statements. The enlargement of the scope of description proceedings was undisputed until the implementation of the IP Enforcement Directive. As a result, it is possible that they could be restricted to the physical reviewing of the premises, machines, devices and processes. Should this be the case, documentary evidence and other evidence on the patent infringement could be obtained through preliminary proceedings.

The review is carried out by the judge (or a court commission), who may be assisted by an independent court expert.

The general principle in Spanish law is that parties must obtain evidence on their own, without interfering in competitors' activities unless this is justified. Accordingly, description proceedings are restricted to cases in which the patentee suspects that there is an infringement but is unable to obtain the evidence needed to file a patent infringement claim. In particular, the patentee should show:

- that a suspected third party is likely to have committed (or is committing) a patent infringement, but that the patent holder lacks the evidence to bring the infringement action immediately; and

- that it is not possible to gather the infringement evidence by other means.

The mere identification of the patent and the potential defendant is insufficient. The patentee must provide the court with facts and legal arguments to support the assertion that the activities carried out by the alleged infringing party may reasonably constitute a patent infringement, and must identify the facts that should be verified by the court (eg the technical characteristics of a given product, or the process or device which is being used by the alleged infringer).

Description proceedings are initiated by a patentee's petition, which fulfil the above requirements. The defendant in these proceedings is the person or the entity that possesses the evidence to be reviewed by the courts in order to establish whether there is an infringement. It is possible that the defendant in description proceedings will not be the same party as the defendant in patent infringement proceedings (eg proceedings against companies that possess or move the potential infringing products).

The court decides on the description proceedings petition in light of the patentee's motion. In addition, before deciding on the petition, the judge may request additional expert reports or order the inquiries that he deems appropriate. Although the latter is theoretically possible, it is rare in practice for judges to order such additional reports and inquiries, and it is therefore advisable that patentees provide sufficient grounds in their petitions.

The decision on the description proceedings petition is taken *ex parte*. The defendant is notified of the decision granting the description proceedings at the moment when they are carried out by the court, or when the court visits the premises to carry out the agreed inquiries. However, in practice, and in order to secure the effectiveness of the inquiries, some courts tend to inform the defendant that an order granting the inquiries has been taken *ex parte*, thus informing him of the date fixed to carry out the agreed inquiries to substantiate facts.

In order to protect the defendant's secret information, the review of processes, devices, products or documents is exclusively carried out by the judge (or by an expressly authorised court clerk) together with a court expert. The patentee is not allowed to be present during the judicial review.

Once the review has been carried out, the expert appointed by the court issues a report or otherwise advises the court of his opinion on the potential infringement. If the judge considers that a patent infringement can be presumed from the evidence obtained through the description proceedings, the judge should, in theory, prepare and draft a report describing the elements used to carry out the potential infringement and should submit this to the patentee. However, in practice, courts tend simply to provide the patentee with the report prepared by the court expert together with a copy of the documents, electronic files, samples or photographs obtained through the description proceedings (usually as attachments to the reports rendered by the court expert).

The reports provided to the patentee as a result of description proceedings can only be used to prepare and file a patent infringement lawsuit, as specified in the relevant petition and only in connection with the facts referred to in such petition.

The patent infringement action must be filed within two months from the date on which the review was conducted, otherwise the evidence obtained through the description proceedings cannot be asserted against the defendant. In any event, if the patentee does not file the action for infringement within this timeframe, it is not prevented from suing the same alleged infringer at a later date in connection with the same patent, but it cannot use the evidence gathered through the description proceedings.

The patent holder is responsible for any losses suffered by the defendant due to the carrying out of the description proceedings. Such losses, however, only include costs and loss of profits directly attributable to the judicial investigation (eg the costs of defendant's employees who assist the judge or the court clerk) Thus, before conducting the description proceedings, the claimant is requested to post a bond in order to guarantee the payment of such potential damages.

In any event, the patent holder may recover the relevant sum through the patent infringement proceedings, as damages consisting of the costs incurred by the claimant to obtain reasonable evidence on the infringement.

(b)   Preliminary proceedings

Preliminary proceedings are addressed to obtain information which is needed to file an action with regard to patent infringement practices that the patentee is aware of (eg cases where the patentee has identified infringing products in the market and knows the identity of the retailer, but has no access to the manufacturer and/or the importer); to define the remedies sought in the infringement proceedings (eg to decide the indemnification criterion or to calculate the hypothetical licence on the basis of the products sold or manufactured by the infringer); or to determine the scope of those remedies (eg to identify the different infringing products regarding which injunctions, withdrawal or compensation orders are sought).

In contrast to description proceedings, as a general rule the patentee is directly provided with the information requested without any court expert deciding whether or not it is likely that the patentee's patent rights are being infringed. Accordingly, in order to avoid the misuse of the proceedings (eg to obtain information on a competitor's suppliers of raw materials) there is a requirement, for the granting of preliminary proceedings, that the patentee must evidence that the requested information is needed to prepare a patent infringement that is likely to have been committed. As an additional measure for the protection of the potential defendant's trade secrets, the judge can act on the defendant's petition and declare the proceedings confidential.

Preliminary proceedings are established for all kinds of litigation. However, specific provisions on preliminary proceedings relating to intellectual property rights proceedings have been introduced as a result of the IP Enforcement Directive. Therefore, in addition to the specific requirement on infringement, the granting of preliminary proceedings is also subject to the fulfilment of the general requirements for preliminary proceedings (ie legitimate interest, justified need and adequacy of the requested information in order to prepare the proceedings).

The decision on the preliminary proceedings is taken *ex parte*, but, in contrast with the description proceedings, the defendant can oppose them before they are conducted. The court decision granting the petition for preliminary proceedings is notified to the defendant in order for it to oppose the patentee's petition on the basis of the lack of fulfilment of the relevant requirements. In the event of opposition, a special hearing is held and the court issues a decision either accepting the opposition and denying the preliminary proceedings, or rejecting it and ordering the conduct of the preliminary proceedings.

In particular, patent holders may be provided with the following information through preliminary proceedings:

- names and addresses of manufacturers, distributors, suppliers and other previous holders of the potentially infringing goods or services, as well as their intended wholesalers and retailers;
- information on quantities manufactured, delivered, received or ordered, as well as the price obtained for the potentially infringing goods or services, and the models (versions) and technical characteristics of such goods and services; and
- banking, financial or commercial documents relating to a certain period of time and under the control of the potential defendant in the infringement proceedings.

Thus, the patentee is entitled to obtain not only the information referred to in Article 8 of the IP Enforcement Directive, but also certain additional information, specifically the models (versions) and technical characteristics of the infringing goods or services.

In the event that the defendant does not provide the court with the agreed information and documents, the court is expressly entitled to order the seizure of such documents.

As in the case of description proceedings, the patent holder is responsible for losses suffered by the defendant due to the implementation of the preliminary proceedings. Therefore, as a prior requirement for the granting of the preliminary proceedings, the patent owner should post a bond to guarantee the payment of such damages.

The information provided to the patentee as a result of preliminary proceedings can only be used to prepare and file a patent infringement lawsuit, as specified in the relevant petition and only in connection with the facts referred to in the petition.

4.2   Proceedings to secure evidence

The purpose of these proceedings is to secure certain forms of evidence that might easily be destroyed after the initiation of the infringement proceeding.

In these proceedings, patentees are not seeking to verify whether a patent infringement is or was being carried out, or to obtain the information needed to file an infringement claim. Their goal is to prevent the evidence relating to a patent infringement from being destroyed. Through these proceedings, patentees ensure that they will be able to rely on the existing evidence to support and prove the infringement. Accordingly, in contrast with description proceedings or preliminary proceedings, the applicant for proceedings to secure evidence can only access the evidence seized in the infringement proceedings.

As with preliminary proceedings, proceedings to secure evidence were not specifically designed for patent or even intellectual property litigation. They were introduced in Spanish law in order to protect any kind of evidence. However, as a result of the implementation of the IP Enforcement Directive, specific provisions for intellectual property litigation have been adopted.

In order to secure evidence, the applicant must be able to show that a third party has probably committed a patent infringement, and that this patent infringement may be evidenced through certain forms of evidence that may easily be destroyed by the third party, by any other party, or by nature. In addition, the patentee must demonstrate that the evidence that is intended to be secured qualifies as acceptable evidence and will be useful within the infringement proceedings. Finally, the applicant must be able to show that the proceedings to secure the evidence could be conducted in a short period of time and without significantly affecting the defendant. Following this logic, proceedings to secure evidence cannot be used as a disguised interim injunction.

In patent litigation, the following measures to secure evidence can be agreed:

- production of samples of infringing products;
- a detailed description of infringing products;
- seizure of potential infringing products; and
- seizure of materials used in the manufacture and distribution of the infringing products.

The proceedings to secure evidence may be applied for either before the filing of the patent infringement claim or during patent infringement proceedings (once the risk of the destruction of evidence arises). If measures to secure evidence are granted prior to the filing of the patent infringement actions, such measures will become ineffective if the applicant does not file a claim within 20 working days following the date on which they were granted.

As a general rule, the court must hear submissions from both parties before issuing a decision on the petition to secure the evidence. However, in the event of potentially irreparable damage, such as the destruction or consumption of the forms of evidence at risk, or as a consequence of the delay due to hearing submissions from the parties, the court may adopt such measures without the defendant having been heard. In such cases, the defendant may oppose the decision to preserve evidence within 20 days of its granting by the court.

4.3   Production of documents

Parties to patent infringement proceedings are not entitled to request the court to order the counterparty to produce all documents potentially referred to the infringement. However, parties are able to ask the court to request the other party to produce certain documents provided that:

- the documents relate to the discussion on the merits,
- they are duly identified by the requesting party; and

- the requesting party evidences that it is unable to access the document.

As a specific requirement in intellectual property litigation in general, and patent litigation in particular, the claimant must provide the court with reasonable evidence that the defendant has infringed the patent right.

There is no list specifying what documents can be requested. However, in patent litigation, the claimant is expressly entitled to request that the defendant produce banking, financial, commercial and importation documents under the control of the infringer. Among others, the parties can request accounting documents (to rebut the damages report prepared by the other party's expert) and the Drug Master File (to rebut the non-infringement report).

When the documents requested by the parties refer to sensitive information, it is possible to request that the court issue a protective order to cover such documents, including an order to keep them confidential and to refrain from exploiting them, or to declare the confidentiality of the proceedings. In addition, it is possible to request that the court does not provide the documents to the party itself, but rather to an appointed independent expert who will prepare the relevant report and maintain the confidentiality of the documents.

5. Availability of interim relief (especially injunctions)

Patent holders or licensees with standing to sue patent infringers are entitled to seek pre-trial relief protection in the event of patent infringement. In addition, any person filing other patent-related actions (ie revocation actions and declaratory actions) is entitled to seek pre-trial relief protection.

Pre-trial relief petitions linked to patent infringement, the most common in Spanish patent litigation, can be sought in the event of either actual or imminent infringement and can be filed against any potential defendant. A very comprehensive set of pre-trial remedies is available to patent right holders and courts do not hesitate to grant robust pre-trial relief to protect rights holders' interests. Accordingly, patent rights holders can be given adequate pre-trial protection in order to minimise (or even avoid) the negative impact of third parties' infringement of their rights, until a decision on the merits is made by the courts.

On a different note, in revocation and declaratory actions, third parties are provided with appropriate/proper pre-trial remedies to avoid patent holders from taking action (eg to assign the patent) which may have a negative impact on the practical effects of the potential ruling revoking the patent or declaring that it has not been infringed vis-à-vis third parties (eg to make a record of the pending revocation proceedings in the Spanish Patent and Trademark Office, or to inform all addressees of infringement warning letters that a revocation action has been filed against the asserted patent). As indicated above, there is no restricted list of potential remedies that can be sought and, therefore, the most appropriate remedies can be requested in light of the circumstances, provided the applicant can persuade the court that they are necessary and proportionate for the protection of the IP rights.

There are three substantive requirements to be met for the granting of pre-trial relief. First, there are the typical requirements for any pre-trial relief petition, namely the likelihood of success (*fumus boni iuris*) and the risk of severe and irreparable prejudice (*periculum in mora*). Secondly, a Spanish feature in patent litigation is that the patentee must evidence that the patented invention is being exploited in a country that is a member of the World Trade Organisation (WTO) and, in addition, that the Spanish market is being sufficiently supplied with the relevant product (eg in the case of pharmaceutical patents for active ingredients, it must be evidenced that the patented active principle is being manufactured in a WTO member state and that the Spanish market is being adequately supplied with the relevant medicine). Thirdly, the applicant must comply with a formal requirement consisting of the offer of a bond to secure the payment of the damages that the defendant may suffer due to the pre-trial relief and must post the bond set by the court as a condition for the enforceability of the interim relief granted.

The *fumus boni iuris* has been construed as the evidence of the ownership of a patent in force and of the *prima facie* infringement of this patent by the defendant. Nullity arguments tend not to be considered by the courts when assessing the *fumus boni iuris* requirement, unless very well grounded and direct arguments are held against the validity of the patent and particularly if the nullity action has been filed before any infringement action has been filed.

The risk that during the conduct of the infringement proceedings the factual situation changes in such a way that would limit the usefulness of any potential decision granting the infringement action has been considered a risk of severe and irreparable damage (*periculum in mora*). Courts tend to consider that the mere risk of starting the imminent infringement or continuing the already committed infringement implies the damage necessary for requesting the granting of interim relief. However, there is no *periculum in mora* when the term between the infringing practice and the filing of the claim is extended. Indeed, according to case law, and provided that the claimant has sufficient evidence to bring the appropriate action, the relevant petition must be filed shortly after the date on which the claimant became aware of the alleged infringement.

Under ordinary circumstances, the pre-trial petition must be filed together with the main claim, in a separate section of the claim writ. Both the pre-trial petition and the main claim are conducted by the same court, but in separate proceedings. Indeed, pre-trial relief proceedings are ancillary proceedings to the main claim, but are dealt with independently.

The filing of a prior, separate injunction is only admitted under exceptional circumstances of duly evidenced urgency. There is no specific provision defining situations of evidenced urgency and it is decided by the courts on a case-by-case basis. Court precedents have construed the urgency requirement for conducting separate pre-trial injunctions in a restrictive manner. If separate pre-trial injunctions are granted, a claim must be filed within 20 working days after the decision granting the injunctions is notified to the parties.

In the relevant petition, the plaintiff must evidence the fulfilment of the requirements for the granting of pre-trial relief with regard to both facts and legal arguments. As a general rule, the possibility for the claimant to produce evidence with regard to pre-trial relief is precluded once the petition has been filed. As an exception to this rule, if the pre-trial remedies sought consist of the cessation of the infringing activities (and this is the case in most patent litigation cases), the claimant is entitled to request the court to order certain inquiries or the production of expert reports that the claimant was unable to produce on its own and that are needed in order to make a decision on the pre-trial relief petition.

The court usually issues a decision on interim injunctions after holding a specific hearing to discuss the fulfilment of the requirements to grant pre-trial relief and to assess the relevant evidence. Indeed, once the petition has been granted, it is served on the defendant and the parties are called to the hearings. During the hearings the defendant orally opposes the granting of the pre-trial relief and submits evidence supporting its allegations (documentary and other evidence is produced and expert witnesses are examined). The claimant's evidence must have been identified in the initial motion for pre-trial relief; whereas the defendant's forms of evidence are identified by the defendant during the hearings. In practice, some courts only allow documentary evidence, including technical reports, to be submitted and reject examination and cross-examination of witnesses and experts.

In addition to the opposition to the granting of pre-trial relief, within the hearings the defendant can request the court to order the substitution of the remedies sought by a bond. If the pre-trial remedy sought as pre-trial relief implies a restriction of manufacturing or commercialising activities of the defendant, the defendant is entitled to request the substitution of the remedy by a bond to secure potential damages arising from the continuation of the infringing actions. The Spanish Supreme Court has clarified that Spanish law does not provide for a compulsory substitution but, rather, establishes the possibility of such substitution if certain requirements are met. Several court precedents have denied the substitution petition on the basis that it would *de facto* hinder the protection pretended by the pre-trial relief petition and on the basis of a strict construction of the necessary restriction or manufacturing or commercial activities.

The pre-trial relief decision addresses not only the remedies granted, but also the substitution petition and the bond, as the case may be. In its decision, the court establishes the amount of the bond to be posted by the plaintiff as a prior requirement for enforcement of the pre-trial relief decision.

In exceptional circumstances, interim injunctions can be decided *ex parte*. In particular, this occurs if the applicant evidences that there is an extremely urgent situation, or that the practical effect of the interim injunction would be hindered if the petition is notified to the defendant, or a hearing is conducted. There is no provision defining such situations and the need for granting *ex parte* pre-trial relief is assessed and decided on a case-by-case basis. Certain precedents have established that an extremely urgent situation is found to exist if a claimant would suffer irreparable damage if no immediate action is taken (due to the time needed to serve the claim on the defendant and to conduct the pre-trial relief proceedings, including the holding of hearings). If interim relief is granted *ex parte*, the relevant decision is immediately enforceable, but the defendant may oppose it once it has been notified of the decision.

The ruling on pre-trial injunctions is not binding upon the court and, therefore, will not affect the final decision to be issued on the merits. The ancillary character of the pre-trial petition implies certain risks that should not be treated lightly. The plaintiff is liable for the damages suffered by the defendant if the pre-trial relief decision is revoked by the second-instance court (regardless of whether or not an infringement is ultimately determined to exist), or if no patent infringement is confirmed by the court at the end of the main proceedings. The plaintiff is subject to a strict liability rule.

6.   Hot topic: product patent protection and TRIPS Agreement in Spain

6.1.   The origin of the controversy

Under previous Spanish patent regulation, the Statute of Industrial Property, which dates back to July 26 1929,[4] the patentability of chemical and pharmaceutical compounds was prohibited. The patent regulations of this law were derogated by the Spanish Patents Act, which is currently in force and which does not include any prohibition on the patentability of chemical and pharmaceutical compounds.

The reason for derogating the former Spanish patent regulations was the Spanish Kingdom's decision to join the European Community. As a requirement of joining the European Community, Spain was compelled to join the European Patent Convention, under which there is no restriction on the patentability of inventions consisting of chemical or pharmaceutical products. Therefore, in order to homogenise both regimes of patent protection and to modernise the Spanish patent system, a new Spanish Patents Act was enacted.

For economic reasons, Spain was interested in deferring the effects of the patentability of chemical and pharmaceutical compounds. Accordingly, two measures were taken.

On the one hand, when joining the European Patent Convention, Spain made a reservation in order to postpone the effects of the patentability of chemical and pharmaceutical compounds in accordance with Article 167(a) of the Convention (the Reservation). The Reservation was made for the maximum term allowed by the Convention, which expired on October 7 1992. In particular, the Reservation reads as follows: "European patents, in so far as they confer protection on chemical and pharmaceutical compounds, shall be ineffective". Therefore, according to the Reservation, product claims for pharmaceutical and chemical compounds of Spanish parts of European patents that were applied for prior to October 7 1992 were unenforceable.

On the other hand, with regard to purely Spanish patents, the patentability of inventions consisting of chemical or pharmaceutical compounds under the Spanish Patents Act was deferred until October 7 1992, by virtue of the First Transitional Provision of the Spanish Patents Act. Accordingly, Spanish patents applied for prior to October 7 1992, in theory, should not protect chemical or pharmaceutical compounds, given that it would imply a breach of the patentability requirements of both the Statute of Industrial Property (applicable to patents applied for until June

---

4   Statute of Industrial Property, dated July 26 1929.

26 1986) and the Spanish Patents Act in accordance with the Transitional Provision (applicable to patents applied for between June 26 1986 and October 7 1992). Therefore, the protection of chemical and pharmaceutical products despite the prohibition constituted a ground for revocation.

The key date for both the Spanish Transitional Provision and the European Reservation is October 7 1992 and therefore patents applied for under their regimes will be in force in Spain until October 7 2012.

If no further patent regulation had entered into force, no conflict would have arisen in relation to these 'old' patents on chemical and pharmaceutical inventions. However, in 1995 Spain joined the World Trade Organisation agreements including the TRIPS Agreement (on Trade-Related Aspects of Intellectual Property Rights), Articles 27 and 70 of which contradict both the content of the Spanish Transitional Provision and the European Reservation.

The impact of TRIPS on the Spanish patent system is a hot topic of discussion in Spain. It is worth noting that it is not a mere theoretical discussion, given that the final position of Spanish courts has significant consequences from an economic point of view.

6.2   The impact of TRIPS on patents applied for prior to October 7 1992

(a)   The TRIPS Agreement: principle of non-discrimination and transitional provisions
Article 27.1 of TRIPS establishes the principle of non-discrimination of inventions according to the field of technology: "... patents shall be available and patent rights enjoyable without discrimination as to the place of the invention, the field of technology and whether products are imported or locally produced".

According to its transitional provisions, the obligations established in the TRIPS Agreement must be complied with in relation to situations that existed prior to the agreement's enforceability. In particular, according to Article 70.2 of TRIPS, all obligations set out in the agreement apply to subject matter that existed and was protected when TRIPS entered into force: "Except as otherwise provided for in this Agreement, this Agreement gives rise to obligations in respect of all subject matter existing at the date of application of this Agreement for the Member in question, and which is protected in that Member on the said date ... ". For the purposes of this provision, "subject matter existing at the date of application of this Agreement", as interpreted by the WTO Dispute Settlement Body in its decision of September 18 2000, comprises both existing patents and the subject matter protected by existing patents.

In addition, Article 70.7 of TRIPS specifically states that patents applied for prior to the entry into force of TRIPS that were pending on that date may be amended to claim any enhanced protection provided under the provisions of TRIPS, subject to the condition that the modification does not include new subject matter:

"In the case of intellectual property rights for which protection is conditional upon registration, applications for protection which are pending on the date of application of this Agreement for the Member in question shall be permitted to be amended to claim any enhanced protection provided under the provisions of this Agreement. Such amendments shall not include new matter."

(b)   **Impact of TRIPS on Spanish Patents Act, the Statute of the Industrial Property and the Reservation**

In Spain, the impact of TRIPS on the patentability of chemical and pharmaceutical compounds has been litigated before the courts during 2007 and 2008. In particular, there has been discussion as to whether TRIPS has superseded (*lex posterior derogat priori*) the Reservation (according to which pharmaceutical or chemical product claims of European patents were not effective in Spain), as well as the provisions of both the Spanish Patents Act and the Statute of Industrial Property (which set out the exception to the patentability of chemical and pharmaceutical compounds).

According to the main case law, since international agreements are directly applicable by the courts in Spain (ie no formal implementation is required), as from the effective date of the TRIPS Agreement in Spain the Spanish authorities are obliged to comply with TRIPS' provisions.

The consequences of the TRIPS Agreement on the patent protection of chemical and pharmaceutical products differ depending on the specific patent at hand. In order to analyse effects of TRIPS, it is necessary to establish groups of cases as follows:

- patents applied for before October 7 1992 and granted before January 1 1996 (the 'old' patents); and
- patents applied for before October 7 1992, but granted on or after January 1 1996.

These groups of patents can be further divided depending on whether or not the patents contain product claims. In addition, some differences arise depending on the kind of patent at hand, either the Spanish part of a European patent or a purely Spanish patent.

(c)   **Old patents**

(i)   *Old patents containing product claims*

There are certain patents that include product claims despite the prohibition. According to the Spanish Patents Act and the Statute of Industrial Property, product claims of Spanish patents which protect chemical or pharmaceutical compounds are null. In addition, pursuant to the Reservation, product claims of Spanish parts of European patents are unenforceable.

There are strong arguments to support the position that, as a result of the combination of Articles 27 and 70.2 of TRIPS, Spain must apply all TRIPS obligations to all subject matter that existed and was protected when TRIPS became applicable in Spain (ie on January 1 1996), including old patents. Indeed, TRIPS has superseded the provisions of the Spanish Patents Act that provided for the nullity of Spanish claims that protected pharmaceutical and chemical compounds. Therefore, as of January 1 1996, Spanish courts may not declare the nullity of these claims. Nevertheless, special rules may apply to patents governed by the Statute of Industrial Property and the impact of TRIPS on them is uncertain.

With regard to Spanish parts of European patents, according to the Reservation, claims that protected pharmaceutical and chemical compounds were unenforceable. In our view, as of January 1 1996, the Reservation is ineffective (it should be considered as implicitly withdrawn) and, therefore, these claims are not affected by any grounds of unenforceability. Several nullity actions were brought against product claims of Spanish parts of European patents on the basis of a breach of the Reservation, which have been rejected by an appellate court ruling. However, there is a first-instance precedent which ruled the nullity of an old patent (the appellate decision relating to this case has not been published yet). In any event, in our view, there are arguments to maintain that this first-instance decision should be revoked.

With regard to Spanish patents, there is no case law dealing specifically with the issue. However, in our opinion there are good arguments to defend the validity of the claims that protected pharmaceutical and chemical compounds of Spanish patents governed by the Spanish Patents Act.

(ii)   *Old patents that do not include product claims*
In our opinion, if a patent was applied for before October 7 1992 and granted before January 1 1996 and does not include product claims, the patent holder is entitled to invoke TRIPS and request the Spanish authorities to add product claims to these patents, taking into account that it was unable to include them for chemical and pharmaceutical compounds on their application date.

In the case of Spanish patents, the patent owner may directly seek the addition of product claims before the Spanish Patent Office by filing the relevant application. In our view, the Spanish Patent Office should grant the modification under Articles 27 and 70.2 of TRIPS. Some petitions have been filed with the Spanish Patent Office and subsequently rejected. Appeals against these decisions have been filed with the Spanish courts for administrative purposes, but the proceedings are still pending.

With regard to Spanish parts of European patents, provided that the claims of the parallel patents for the contracting states that did not make a reservation contain product claims, it should be possible to file a new translation into Spanish of the claims of the patent which includes the product claims. Although in our opinion this argument is well grounded, the Spanish Patent Office has rejected the petitions to file new translations. Judicial appeals against these decisions have also been filed, but no decision has yet been made. It is worth mentioning that a first-instance ruling issued by a commercial court has found that, notwithstanding the fact that the reviewed translation of a European patent including product claims was rejected by the Spanish Patent Office and proceedings were pending against this decision, the defendant had infringed those product claims, and appropriate remedies were granted.

(d)   ***Patents applied for before October 7 1992 and granted on or after January 1 1996***
Article 70.7 specifically states that patents applied for before October 7 1992 that were pending on the date of application of the agreement (ie patents that were granted on or after January 1 1996) will be allowed to be amended in order to claim any enhanced protection provided under TRIPS, subject to the condition that they do not include new claims.

As a consequence, any patent including product claims is valid and enforceable. On the contrary, if the patent does not contain any product claim, the Spanish

Patent and Trademark Office has rejected petitions to add product claims on the grounds that the patent holder had the right to request the addition but did not assert that right within the period allowed. The decisions of the Spanish Patent Office rejecting the said petitions have been appealed to the administrative courts and the relevant proceedings are currently (late 2008) pending.

On the other hand, there are several precedents affirming the validity of product claims of European patents applied for before TRIPS entered into force and subsequently granted even where there was no formal amendment of the patent (ie the patent application included product claims as filed and, therefore, no modification was carried out). Although there is no precedent relating to Spanish patents, the reasoning of the existing precedents should apply and, in our opinion, the validity of Spanish product claims for chemical and pharmaceutical compounds would also be upheld.

United Kingdom

Paul England
Sebastian Moore
Herbert Smith LLP

UK patent litigation can be conducted before the patent courts of England and Wales or the Scottish courts. In Scotland, intellectual property disputes are normally raised in the Court of Sessions and there is no specialist court for intellectual property cases. In practice, few cases are brought before the Scottish courts and most UK patent litigation takes place in the patents courts. For this reason the following chapter concerns the practice and procedure of the patents courts only.

1. Lawyers

The team of attorneys involved in litigation is split between a firm of solicitors, who directly represent the client, and barrister(s) (known as counsel). The solicitor is the point of contact for the client and is responsible for the management of the litigation and the preparation of evidence. Counsel are instructed by the solicitors to assist in the drafting of formal court documents such as statements of case (eg particulars of claim, defence and so on) and skeleton arguments, and also to undertake the advocacy at trial, although this may also be carried out by a solicitor who has carried out the necessary training to become a 'solicitor advocate'. Counsel themselves are split into two types:

- barristers who are perceived to be at the top of their profession and have been awarded the title 'Queen's Counsel' by an independent panel; and
- all other barristers, who are known as 'Junior Counsel'.

Junior Counsel generally draft the statements of case, undertakes the review of evidence (alongside the solicitors) and prepares the case for trial. Queen's Counsel will usually undertake the advocacy work in court and oversee the tactics and strategy used in the litigation (again, along with the solicitors). Counsel will often be involved throughout the preparation of the case from the start of proceedings until trial and appeal.

Occasionally, in complex cases, a patent attorney may form part of the team in order to advise on technical matters such as drafting amendments and prosecution history.

2. The court system

2.1 Courts of first instance

There is generally little scope for forum shopping within England and Wales on patent matters. This is because there are only two courts which have exclusive jurisdiction over patent matters:

- the Patents Court of the Chancery Division of the High Court; and
- the Patents County Court.

Both the High Court and the Patents County Court are situated in central London, but it is possible for them to sit outside London on application. These courts can hear a whole range of disputes relating to patents, including, revocation, infringement, declarations of non-infringement, entitlement, threats, employee compensation and licence disputes, as well as post-grant amendment.

There is no formal division of jurisdiction between the Patents County Court and the Patents Court, so the County Court can in theory hear cases of unlimited complexity and value. There is also no restriction on the levels of costs the Patents County Court can award. However, it is generally the case that actions of lower value or complexity are in fact dealt with by the Patents County Court, and the more complex and high value cases dealt with by the Patents Court. This is facilitated to some extent by the transfer of cases between the two courts in appropriate cases.

In addition to the court system in England and Wales, a person may also apply to the Intellectual Property Office (IPO) for the determination of patent-related disputes. The IPO has the power to make decisions on:

- the ownership of a patent (ie entitlement, co-ownership issues, inventorship and employee compensation);
- issues regarding terms of licences and compulsory licensing; and
- the validity of a patent (ie revocation and amendment).

Furthermore, the IPO can deal with the infringement of a patent provided that both the patentee and the alleged infringer agree that the matter should be referred to the IPO. Proceedings in the IPO are separate from those in the courts and are subject to their own procedures. In particular, the proportion of legal costs recoverable by the winning party from the losing party is much smaller in cases before the IPO.

The IPO also has a procedure for issuing opinions whereby anyone can request an opinion on questions of infringement and validity of a UK patent. The opinions are prepared by an examiner and are not binding for any purposes. The procedure was introduced as a means of providing a quick and affordable way for patentees to obtain an assessment of validity/infringement with a view to encouraging settlement. The IPO also offers a mediation service as a further alternative method of dispute resolution to litigation.

2.2   Appeals

There is no automatic right of appeal in the United Kingdom. Instead, permission must be given either by the judge of the court that heard the case at first instance, be it the Patents County Court or the Patent Court, or by the Court of Appeal. The application for permission to appeal is made initially to the judge of the first-instance court. If the judge refuses to grant permission, an application is made on paper to the

Court of Appeal followed by an oral hearing if it is again unsuccessful on paper. The application must explain why the applicant considers that the appeal has a real prospect of success. A 'real prospect of success' means that the appeal has a realistic, as opposed to fanciful, prospect of success. Alternatively, an appeal may be pursued when there is some other compelling reason why the appeal should be heard. Most complex patent cases are given permission to appeal (by either the trial judge or the Court of Appeal) and the approach that the Court of Appeal now takes is to grant permission unless it is readily apparent from the papers that an appeal is unlikely to succeed. If successful, the appeal will then be heard by the Court of Appeal.

Appeals in England and Wales are generally on discrete points of principle (ie application of the law) and there is no retrial of all the issues as sometimes occurs in civil law jurisdictions. Appeals are based on alleged mistakes of points of law or misinterpretation of evidence by the judge of first instance. Only in very limited circumstances can fresh evidence be adduced in an appeal.

Appeals from patent decisions of the IPO are made to the Patents Court. Appeals of opinions of the IPO may also be made to the Patents Court. Whilst the appealed opinion will remain non-binding, it may be perceived as carrying greater weight if upheld on appeal.

In exceptional cases, it is possible for either party to make a further appeal from the Court of Appeal to the highest appeal court in the United Kingdom, known as the Appellate Committee of the House of Lords, or the 'House of Lords' for short. The party wishing to appeal must first obtain permission from the House of Lords. Such leave is rarely given as the decision being appealed must be based on an important issue of law that requires clarification, or is of public interest.

In certain cases an appeal may be made directly from the Patents Court to the House of Lords, cutting out the Court of Appeal. This is known as a 'leapfrog' appeal. Before a leapfrog appeal can proceed a certificate must first be obtained from the Patents Court, subject to satisfying a number of requirements, followed by leave of the House of Lords.

2.3 The patents judges

The judges of the Patents Court and the Patents County Court are intellectual property law specialists. Many of them practised at the intellectual property bar for years before becoming judges. Currently, four of the patents judges have scientific backgrounds.

Generally, patent appeals are heard by three Court of Appeal judges, one of whom will previously have been a first-instance Patents Court judge. The judges of the House of Lords are all ex-Court of Appeal judges. These judges are called Lords of Appeal in Ordinary. Five judges in the House of Lords will hear each case although, exceptionally, more than this may sit. In both the Court of Appeal and the House of Lords, dissenting opinions may be given but the decision will be determined by the majority.

2.4 Parallel decisions

Patent matters in the United Kingdom are governed primarily by the Patents Act

1977, as amended by the Patents Act 2004 (together hereafter the Patents Act). This enacts into English law many of the provisions of the European Patents Convention (EPC) and the 2000 European Patents Convention (EPC 2000). The courts will be mindful of parallel decisions made in countries that are signatories to these instruments. More influential are decisions of the European Patent Office (EPO). As Lord Hoffmann pointed out in *Merrell Dow Pharmaceuticals v Norton* [1996] RPC 76, UK courts would normally follow settled EPO jurisprudence:

> *"... the United Kingdom Courts ... must have regard to the decisions of the European Patent Office ('EPO') on the construction of the EPC . These decisions are not strictly binding upon courts in the United Kingdom but they are of great persuasive authority."*

However, the courts are not bound to take such other decisions into account and they will appreciate that because of procedural and evidential differences the opportunities for harmonising decisions can be limited. In particular, as regards the EPO, Lord Justice Jacob has recently stated in *Actavis UK Limited v Merck & Co Inc* [2008] RPC 26 that whilst our courts would and should normally follow the settled jurisprudence of the EPO, it should be understood that they are not bound to do so. He also added:

> *"In the unlikely event that we are convinced that the commodore [the EPO] is steering the convoy towards the rocks we can steer our ship [the UK courts] away."*

Patent law decisions made in the higher courts of other common-law jurisdictions, such as Canada and Australia, can also be informative for the court, and to a lesser extent decisions made in the United States.

3.   Procedure and timescale of proceedings

3.1   Pre-action issues

There is no pre-action protocol specific to patent proceedings that must be followed prior to starting proceedings. However, there is a general practice direction contained within the UK's Civil Procedure Rules that govern proceedings in England and Wales as a whole. This states that parties should act reasonably in exchanging information and documents relevant to the claim and generally in trying to avoid the necessity for the start of proceedings. This requires that the potential claimant party writes to the potential defendant with details of the claim and the defendant responds on these matters in a reasonable time. It also envisages the parties negotiating to settle the claim 'economically' and without court proceedings. However, in high-value patents cases in which the parties are both sophisticated organisations who can afford legal advice, this practice direction often appears to be disregarded without criticism or costs penalties being imposed by the court.

Moreover, in cases of alleged infringement, warning letters may actually give rise to liability if they contain groundless threats of infringement proceedings. The threats provisions in the Patents Act are complex and must be consulted carefully before drafting a warning letter. Remedies are available to an aggrieved party to whom groundless threats of proceedings for infringement of patents and patent applications are made, and such a party may bring proceedings against the party making these threats. The types of relief available include an injunction to prevent further threats, a declaration that the threats are unjustifiable and/or damages in respect of any loss sustained.

The claimant is not entitled to relief under the threats provisions, even if the patent is found invalid in a relevant respect, if the defendant proves that at the time of making the threats he did not know and had no reason to suspect that the patent was invalid in that respect.

Finally, solicitors and patent agents are at risk of a threats action if they write a 'letter before action' (even if they are writing on behalf of a client) that is outside the scope of the above permitted acts.

3.2 Standard procedure

Proceedings are started by issuing and serving a claim form. This may be served with particulars of claim, or the particulars of claim may follow within the time limit (see the diagram overleaf). If infringement is alleged, the particulars of claim should be accompanied by a document known as the "Particulars of Infringement", which gives details of alleged infringing acts. If invalidity of a patent is claimed, the particulars of claim must be accompanied by a document called the "Grounds of Invalidity", which gives details of the alleged invalidity. Where the claim contains allegations of patent infringement, the defendant has within 42 days from service of the claim form to file his defence and/or counterclaim. If the complaint merely concerns the revocation of a patent, the patentee has 28 days after serving the particulars of claim and grounds of invalidity to file his defence and/or counterclaim (provided that an acknowledgment of service of those particulars of claim has been filed within 14 days). However, it may be possible to extend the time for filing a defence by obtaining the consent of the counterparties, or by applying to the court. In practice, an extension of up to 28 days is often requested and granted.

The defence must set out the defendant's case against the claims made. The defendant to an infringement action may plead invalidity as a defence and counterclaim to revoke the patent. If both infringement and validity are at issue they will generally be heard together by the court, although a party may ask the court to determine the issue of infringement as a preliminary issue. However, this is unusual. This is because, as Mr Justice Laddie stated in *Coin Controls Ltd v Suzo International (UK) Ltd* [1998] 3 WLR 420, infringement and validity of an intellectual property right "are so closely interrelated that they should be treated … as one issue or claim".

The stages in typical patent proceedings are shown in the diagram overleaf. They may comprise a number of steps, including a case management conference (a hearing at which the future timetable of the proceedings is directed by the court), disclosure of documents (including a product/process description when infringement is at issue and a schedule of commercial success, if relevant), experiments, naming of experts and exchange of expert and witness evidence. First-instance infringement proceedings take about one year from issuing proceedings to judgment.

An application for leave to appeal must be made within 21 days from the date of judgment and is typically dealt with in a few weeks. A decision of the Court of Appeal will take approximately 12 months from this time. An appeal to the House of Lords will currently take approximately 18 months to result in a decision.

Issue and serve claim form (and particulars) (and grounds of invalidity)	The claim form must be served within four months of being issued. The particulars need not be served at the same time (see below). Grounds of invalidity are served only when validity is challenged.
Service of the particulars of claim (and grounds of invalidity)	Within 14 days of service of the claim form if not served with the claim form.
File and serve defence	28 days after service of the particulars of claim if claim concerns validity issues only (if acknowledgement of service filed). Within 42 days of service of the claim form if infringement is alleged.
File and serve reply	Within 21 days of service of the defence.
Apply for case management conference	Within 14 days of service of defence.
Case management conference	Typically heard about three months from claim, depending on the court timetable and the availability of the parties.
Disclosure and inspection	Typically about six months after service of the claim, depending on the number of documents at issue and their complexity.
Experiments	If applicable, any time from disclosure to trial.
Exchange of evidence	Typically, approximately ten months from service of the claim.
Evidence in reply	Typically one month after exchange of evidence.
Trial of infringement and validity issues)	Approximately 12 months from claim.
Appeal	Approximately 24 months from claim.

3.3 Streamlined procedure

In appropriate cases, the Patents Court also offers a streamlined procedure. If agreed by all parties to the litigation or if ordered by the court, more simple patent cases can be allocated to this procedure. Under the streamlined procedure, trial will not normally last more than one day, all evidence is in writing (including expert evidence), there is no experimental evidence, and there is no requirement to give disclosure. The advantage is that once this procedure has been ordered or agreed to be used by the parties, then the trial will generally be heard within six months of service of the claim form.

3.4 Stay pending EPO proceedings

The patents courts have a discretion whether to grant a stay of revocation proceedings when parallel opposition proceedings are afoot in the EPO. Until recently, it had been thought that there was a presumption in favour of such a stay. However, the recent Court of Appeal decision in *Glaxo Group Ltd v (1) Genentech and (2) Biogen* [2008] FSR 18 suggests that this is not the case. Instead, the length of the proposed stay will normally be the most significant factor in deciding whether it is appropriate to stay UK proceedings. Given the slow pace of opposition proceedings in the EPO, it will almost always be the case that UK proceedings will be concluded much earlier. It will therefore be difficult in future to obtain a stay of the UK proceedings, although applications for stays of this kind are always decided on a case-by-case basis and depend very much on the facts at issue.

4. Availability of pre-action evidence gathering/disclosure

4.1 Pre-action disclosure

A party can request pre-action disclosure from a prospective counterparty for the purpose of gathering evidence to support a claim of patent infringement. It is a matter for the court's discretion whether such disclosure will be ordered. Requests for pre-action disclosure must be focused, with particular categories of documents being identified; the courts will not allow a party to 'go fishing' for documents which it does not have reason to believe exist, or which are not relevant to the matters likely to be in issue.

An application for pre-action disclosure must be supported by evidence in the form of a witness statement which satisfies the court that:
- the applicant and the respondent from whom disclosure is sought are likely to be parties to the subsequent proceedings;
- the documents which are sought to be disclosed must be of a type which would be disclosable in the contemplated proceedings had they already commenced; and
- disclosure of such documents before proceedings commence is desirable to dispose fairly of the anticipated proceedings, or to encourage settlement or to save costs.

Documents disclosed under this procedure may only be used for the purpose of the proceedings for which they were disclosed, unless the court gives permission for, or the parties agree to, another use. There will also be no restriction on the use of documents that have been read or referred to in a hearing held in public.

4.2   Disclosure from third parties

In certain circumstances, an order may be made against a third party so as to assist a claimant in gathering information about an alleged wrongdoer, for example their identity, so that a case may be prepared against them. The principle for this is stated in the leading case of *Norwich Pharmacal v Customs and Excise Commissioners* [1974] AC 133:

> "*if through no fault of his own a person gets mixed up in the tortious acts of others so as to facilitate their wrongdoing he may incur no personal liability but he comes under a duty to assist the person who has been wronged by giving him full information and disclosing the identity of the wrongdoer*".

However, jurisdiction under the *Norwich Pharmacal* principle is not limited to ordering the disclosure of a wrongdoer and has broader application to obtain information from a third party who has unwittingly become involved in wrongful conduct that has infringed the rights of a claimant or potential claimant.

4.3   Search and seizure orders

Search orders are often referred to as *Anton Piller* orders after the case in which they first came to prominence, but they are now largely governed by the Civil Procedure Rules (CPR). They are used to enable one party to search the property of another and seize evidence, without warning. They are particularly applicable to cases of counterfeiting where prior notice of the application might result in the destruction of the alleged counterfeit goods.

However, because of their draconian nature, they are difficult to obtain and are uncommon. To obtain such an order there must be:

- an extremely strong *prima facie* case against the respondent;
- the damage, potential or actual, must be very serious for the applicant; and
- there must be clear evidence that the potential defendant has in his possession documents or things that would give rise to liability and that there is a real possibility that this material would be destroyed before an *inter partes* application can be made.

If there is a risk that the applicant will not be able to meet the respondent's costs if the order turns out to be wrongfully made, the court will require a bank guarantee or other security before making the order.

5.   Availability of interim relief

Urgent matters can be dealt with very quickly by the courts. A party can apply for an interim injunction (in cases of extreme urgency, *ex parte*) on very short notice and obtain an injunction almost immediately, at the court's discretion. Such a procedure may be suitable in circumstances where the patentee becomes aware that infringing goods are about to be put on the market in the United Kingdom and wishes to prevent the infringer from launching its product. However, any delay on the part of the patentee in bringing the application for an injunction will greatly diminish the chances of success.

Less urgent matters are dealt with at a hearing attended by both parties (*inter partes*). In these circumstances an injunction can be obtained usually within a couple of weeks, again at the court's discretion. However, again if the patentee delays significantly in making the application, then the court is unlikely to be convinced that there is sufficient urgency to warrant injunction and that the matter cannot wait until a full trial on the merits.

In deciding whether to grant a preliminary injunction, the court will give primary regard to the balance of convenience between the parties of the injunction, provided that the applicant has made out that there is a serious issue to be tried. The court will not indulge in a full trial on the merits of the action any further than it needs to test this latter requirement. The balance of convenience refers to the relative commercial impact that an injunction (or its refusal) would likely have on both parties. A key issue is whether the patentee applicant could be adequately compensated by damages for the alleged infringement if it were successful at trial. If they could be so compensated, then the injunction is unlikely to be given. Determining this will depend on the difficulty of quantifying the applicant's loss. For example, factors the court may consider in the patentee's favour include whether the allegedly infringing product causes permanent price erosion to the applicant's product or would lose the foothold they have obtained in the market for the patented product. Similarly, the court may be in favour of an injunction if the patentee's product is one that will have a short commercial life-span, such that much of the benefit of its monopoly will be damaged by competition from the alleged infringer before trial. However, where the balance of convenience is evenly balanced, the court will look to maintaining the status quo.

If the injunction is granted, the patentee will need to give an undertaking to compensate the alleged infringer for any damage suffered in consequence of the injunction should it be decided at the infringement proceedings to have been wrongly granted (eg because there is no infringement or the patent is found to be invalid). The ability of the patentee to make good this undertaking may also be a factor the court considers in deciding whether to grant the injunction. In some cases, the undertaking may need to be supported by a bank guarantee or a payment into court.

Applications for interim injunctions in patent matters are heard by the specialist patent judges of the Patents Court division of the High Court. This is the same forum that deals with infringement and validity. However, in exceptional circumstances (eg in very urgent matters) an *ex parte* injunction application may be heard by a non-specialist judge of the High Court.

Application for an interim injunction is not without risk. The risk that the injunction will be overturned at trial, exposing the patentee to damages under the undertaking, is discussed above. If the issues on infringement are likely to be less involved compared with the validity proceedings, an alternative to a pre-trial injunction may be to commence infringement proceedings and press for an expedited hearing under the streamlined procedure described above. If successful at trial, the patentee would then be able to obtain a final injunction.

6. Disclosure

Unlike other European jurisdictions, which have much narrower provisions relating to discovery, the normal rule in the patents courts is that all relevant non-privileged documents must be disclosed by each party. Disclosure, and subsequent inspection of documents, will normally be ordered at the case management conference to take place at a set date. However, in cases run according to the streamlined procedure where disclosure is considered to be of little assistance, it may not be ordered at all.

For issues relating to validity, disclosure of documents is limited to the period two years before the earliest claimed priority date and two years after that date. For issues concerning infringement, disclosure of documents is not required if the alleged infringer has served full particulars of the allegedly infringing product or process in the form of a product/process description. For all other issues, standard disclosure is required.

Where the patentee raises the commercial success of its product/process in response to an attack of obviousness, the patentee must serve a schedule containing an identification of the product/process, a summary of sales/revenue received from the product/process, a summary of sales/revenue received for an equivalent prior product/process for equivalent periods of sale, and a summary of any expenditure on advertising and promotion which supported the marketing of the product/process.

Furthermore, parties are also able to apply to the court for an order of specific disclosure. Such an order would compel a party to disclose certain documents or classes of documents, carry out a search or disclose any documents located as a result of a search.

Where disclosure is required, a party must disclose all documents:
- which he relies on; and
- which adversely affect his case, adversely affect the other party's case, or support the other party's case.

However, the party need only undertake a reasonable search for these documents. In determining the reasonableness of the search, the party can take into account the number of documents, the nature and complexity of the proceedings, the ease and expense of retrieval of documents, and the significance of any document that is likely to be located in the search.

However, there are a number of grounds that allow a party to withhold inspection of certain documents:
- The most commonly used ground is legal advice privilege – this protects communications between a lawyer and his client created for the purpose of giving or receiving legal advice, or documents evidencing the content of such communications (where proceedings are not reasonably in prospect);
- A further type of privilege – litigation privilege – protects documents which were created for the dominant purpose of gathering evidence for use in proceedings (or proceedings that were reasonably in prospect) and for giving

legal advice in relation to such proceedings; and

- Without prejudice communications which are made with the intention of seeking a settlement of litigation are also privileged from disclosure. It should be noted that it is the substance of the communication that determines whether it is without prejudice, not whether or not it is labelled 'without prejudice'.

There is no exception to the obligations of disclosure based on trade secrets, although it is advisable to put in place a confidentiality agreement between the parties to the litigation that restricts the disclosure of sensitive documents to named individuals. If the court considers it appropriate, it may hear issues relating to confidential information in private (*in camera*) and make an order protecting the confidentiality of certain information after trial.

The CPR state that documents disclosed in proceedings may only be used for those proceedings, subject to the agreement of the parties or an order of the court to the contrary. Hence, the normal position is that any document obtained via UK disclosure cannot be used in proceedings elsewhere, unless and until it is read in open court in the proceedings in which it was disclosed.

7. Evidence

Together with the disclosure process, the manner in which evidence is dealt with in English procedure can enable a very thorough analysis of the case, making the courts of England and Wales an excellent forum for determining and testing facts. Consequently, a great deal of time and effort is devoted to preparing very detailed evidence of experts on issues relating to validity and infringement. This may be supplemented by more limited evidence from witnesses of fact. The evidence in chief is presented in written form (ie expert reports and/or witness statements) and the witnesses/experts are generally called at trial and cross-examined on this evidence.

7.1 Experts

In English proceedings, technical experts are often central to patent cases. It is therefore very important to devote time to select an expert who is credible and supports your case. Each side will normally engage their own expert(s), to whom they will pay such fee as the expert requires; the number of experts appointed will depend on the nature of the case, but it is rare to have more than two or three per party. In exceptional cases, the court may also appoint an expert. The role of the expert is to assist the court with matters within his expertise (and which are outside the judge's expertise, that is, on technical aspects of the case).

The expert has an overriding duty to the court rather than to the party that has appointed and pays him. In particular, the report prepared by the expert has to represent his true opinion. It is important that the expert complies with his duty to the court at all times and he must make a statement to the effect that he understands duties to the court in his expert reports. It is therefore crucial to find an expert who truly supports your case, especially since this will be tested in cross-examination.

Expert opinion is primarily adduced by way of a written expert report. This is usually exchanged with the other party approximately two months before trial. An expert's report should be prepared by the expert and must be addressed to the court. The content of the report will contain details of the expert's qualifications, the material he has relied on in making the report, the substance of the facts and instructions given to the expert, his conclusions and a signed statement that the expert understands his duty to the court and that he will continue to comply with that duty. In essence those duties are:

- An expert's job is to help the court decide issues by offering his objective and unbiased opinion on matters within his expertise. He should never assume the role of an advocate;
- An expert witness should state the facts or assumptions on which his opinion is based. He should not omit consideration of material facts which detract from his concluded opinion; and
- An expert witness should make the court aware when a particular question or issue falls outside of his experience.

At trial, the expert report is admitted as evidence in chief without any examination. The other party does, however, have the chance to challenge the other side's expert during trial by oral cross-examination of the expert on oath in the witness box.

7.2 Witnesses of fact

Evidence may also be brought from witnesses of fact. In infringement cases these are more likely to be officers of a party who have first-hand knowledge of infringing acts, where these are under dispute. In validity cases, particularly those where inventiveness is at issue, the inventor may give evidence of how they arrived at the invention and its background. However, it should be remembered that such evidence from the inventor is of very limited value in English patent proceedings compared with the evidence of experts.

7.3 Experiments

Experiments are often conducted in English patent proceedings, usually to attempt to prove infringement or matters of enablement for novelty and sufficiency purposes. More rarely, they are used to attempt to prove lack of inventive step. The parties need leave of the court to rely on experimental evidence. Experiments are normally ordered at the behest of one party at the case management conference stage. The order will allow for the party conducting the experiment to serve on the other party a notice of experiments on a particular date. The notice must provide details of the experiment undertaken and the results. The other party may then accept this notice without challenge, in which case the results will stand as evidence at trial. Alternatively, the other party may ask to see a repeat of the experiment, which they will attend to inspect the conduct of the experiment from start to finish. If a repeat is conducted, it is the results of the repeat that are the primary evidence of the experiments at trial. Note that once a notice of experiments has been served, the other party is entitled to disclosure of all the background documents, such as lab notebooks, that relate to the conduct of the experiment at issue, including the results of any work-up experiments.

Experiments usually add a substantial amount of time and costs to the proceedings, so the parties should think carefully before requesting leave to rely on experimental evidence.

8. Law

8.1 Infringement

The infringing acts in the United Kingdom are drawn from the Community Patent Convention (itself not in force). Acts of infringement include: making, disposing of, offering to dispose of, using or keeping a product; and using or offering the use of a process in the United Kingdom, when it is known, or obvious to a reasonable person in the circumstances, that its use there without the proprietor's consent would be an infringement. There are similar provisions for acts with a product obtained directly by means of an infringing process. Furthermore, it is also an infringement to supply or offer to supply in the United Kingdom, without authorisation from the patentee, a person who is not a licensee or otherwise entitled to work the invention with any of the means, relating to an essential element of the invention, for putting the invention into effect in the United Kingdom. This latter infringement also requires that the person supplying knows, or it is obvious to a reasonable person in the circumstances, that those means are suitable for putting, and are intended to put, the invention into effect in the United Kingdom.

Direct infringement of a patent occurs under English law when a product or process, other than that of the patentee, contains all the essential features of the claim (the integers), such that it would anticipate the claim had it been made or performed before the priority date of the patent. This, of course, requires that the patent claims are first construed. Like other signatory states to the EPC, the United Kingdom has implemented article 69 into national law. This states that the extent of the protection conferred by a patent is to be determined by the terms of the claims, and that the description and drawings shall be used in the interpretation of the claims. Likewise, the United Kingdom has implemented the Protocol on the Interpretation of Article 69 of the EPC, including article 2 of the protocol, added by EPC 2000:

> "For the purpose of determining the extent of protection conferred by a European patent, due account shall be taken of any element which is equivalent to an element specified in the claims."

However, the new wording of article 2 of the protocol has not been interpreted as introducing a doctrine of equivalents into English patent law that can take the invention outside the scope of the claims. This was made clear in the leading authority on the issue of claim construction, namely the House of Lords decision in *Kirin-Amgen v Hoechst Marion Roussel* [2004] UKHL 46, although Lord Hoffmann did state that: "there is no reason why equivalence cannot be an important part of the background of facts known to the skilled man which would affect what he understood the claims to mean".

In *Kirin-Amgen*, the House of Lords set out the correct approach to claim construction, mindful of what the protocol forbids, *per* Lord Hoffmann:

> *"The Protocol, as I have said, is a Protocol for the construction of article 69 and does not expressly lay down any principle for the constructions of claims. It does say what principle should not be followed, namely the old English literalism, but otherwise it says only that one should not go outside the claims. It does however say that the object is to combine a fair protection for the patentee with a reasonable degree of certainty for third parties. How is this to be achieved? The claims must be construed in a way which attempts, so far as is possible in an imperfect world, not to disappoint the reasonable expectations of either side. What principle of interpretation would give fair protection to the patentee? Surely, a principle which would give him the full extent of the monopoly which the person skilled in the art would think he was intending to claim. And what principle would provide a reasonable degree of protection for third parties? Surely again, a principle which would not give the patentee more than the full extent of the monopoly which the person skilled in the art would think that he was intending to claim. Indeed, any other principle would also be unfair to the patentee, because it would unreasonably expose the patent to claims of invalidity on grounds of anticipation or insufficiency."*

In short, therefore, construction is a matter for the court to determine what the claims would mean to a skilled man – what would the skilled reader of the patent understand the patentee to have meant by the wording of the claims? Note that this is not the same as determining what the patentee meant when drafting the claims – the test is an objective assessment of what the claims are intended to cover.

8.2 Novelty

Novelty is dealt with in Section 2 of the Patents Act. This states in subsection (1) that an invention will be new if it does not form part of the state of the art. Subsection (2) goes on to state that the state of the art comprises all matter (whether a product, a process, information about either, or anything else) which has at any time before the priority date of that invention been made available to the public (whether in the United Kingdom or elsewhere) by written or oral description, by use or in any other way. This reflects similar wording in article 54(2) of the EPC.

The requirement under the Patents Act that a prior product or prior use must have been made available to the public is a relatively low hurdle. Communication of the prior matter to one other person is sufficient to become 'public knowledge', provided that the person is free in law and equity to use it (ie disclosure has not been made under an obligation of confidence). Any document is regarded as having been published, and therefore forms part of the state of the art, if it can be inspected as of right by the public, whether on payment of a fee or not.

The leading case in the United Kingdom in relation to novelty is *Synthon BV v Smithkline Beecham plc* [2005] UKHL 59, a decision by the House of Lords. The Lords held that, for an invention to be anticipated, the matter relied on as prior art must:

- disclose the claimed invention (disclosure); and
- together with the common general knowledge, enable the ordinary skilled person to perform the invention (enablement).

When considering disclosure under the *Synthon* test, the prior disclosure must be construed as it would have been understood by a skilled person with the benefit of the common general knowledge at the priority date of the patent in suit and not with the benefit of hindsight. It must disclose the subject matter such that, if the prior invention was performed, it would necessarily result (ie more than a possibility or a likely consequence) in an infringement of the patent.

The test for enablement of an anticipating disclosure is the same as the test for enablement for the purposes of sufficiency. Thus, the ordinary skilled person would have been able to perform the invention that satisfies the requirement of disclosure without undue effort or burden, in considering which the skilled person is assumed to be willing to make trial-and-error experiments to get the invention to work.

8.3 Inventive step

Articles 52(1) and 56 of the EPC, corresponding to Sections 1(1)(b) and 3 of the Patents Act, provide that an invention shall be taken to involve a patentable inventive step if it is not obvious to a person skilled in the art, having regard to any matter which forms part of the state of the art. The leading authority on the approach to inventive step (whether the claim is obvious) is now *Pozzoli SPA v BDMO SA and others* [2007] EWCA Civ 588, restating and amending the classic test in *Windsurfing International Inc v Tabur Marine* [1985] RPC 59. The test is as follows:

- identify the notional person skilled in the art and the relevant common general knowledge of that person;
- identify the inventive concept of the claim in question or, if that cannot readily be done, construe the claim;
- identify what, if any, differences exist between the matter cited as forming part of the state of the art and the inventive concept of the claim or the claim as construed; and
- decide whether those differences, viewed without any knowledge of the alleged invention as claimed, constitute steps which would have been obvious to the person skilled in the art, or whether they require any degree of invention.

There are a number of different factors that the court might take into account in making a substantive assessment of obviousness, although the courts make it clear that these are not a substitute for the statutory test itself. The approach taken will depend upon the particular facts and circumstances of the case under consideration. These include:

- *Right to work* An invention will be regarded as obvious if a claim to it would inhibit the rights of a skilled workman to carry out routine modifications of what is already in the public domain. The leading authority for this is *Windsurfing International Inc v Tabur Marine (Great Britain) Ltd* [1985] RPC 59;
- *Why was it not done before?* Whilst the fact that nobody has followed a particular path before does not dispose of an obviousness objection (otherwise any invention which was new would automatically be inventive), the reasons why this has not been done before can be important. If the inventor has solved a long-recognised problem by means which others could

have used but did not, then there may be an inventive step. The leading case on this is *Minnesota Mining and Manufacturing Co v Rennicks (UK) Limited* [1992] RPC 331;

- *Fulfilling a need* Evidence that an invention fulfils a long-felt want and has been commercially successful may be taken into account in assessing obviousness. The leading case is *Haberman v Jackal* [1999] FSR 685;

- *Obvious to try* An invention will be obvious if a person skilled in the art would assess the likelihood of success sufficient to warrant a try. The UK House of Lords has recently considered this test in *Conor Medsystems Inc v Angiotech Pharmaceuticals Inc* [2008] UKHL 49. The court stated that the obvious to try test was only useful in a situation where there was a fair expectation of success and that this would depend on the facts of the case;

- *Overcoming a technical prejudice* An invention may be regarded as inventive if it goes against the generally accepted views and practices in the art. It must be clear that the technical prejudice which the inventor claims to have overcome did in fact exist, and that it was not justified; and

- *Teaching away* This refers to common general knowledge or prior art that acts as a technical prejudice for taking a certain route. In *Dyson v Hoover* [2002] RPC 22, the Court of Appeal stated that common general knowledge has positive *and* negative aspects and that the knowledge of the skilled person should be imbued with both types. In *Dyson*, a case concerning the new technology of bagless cyclone suction for vacuum cleaners, the judges said:
  "such was the 'mindset' within the vacuum cleaner industry, no notional, right-thinking addressee would ever have considered the viability of purifying dirt-laden air from a vacuum cleaning operation, other than by means using a bag [alone] or bag and final filter ... this negative thinking which as [counsel] suggested amounted to prejudice, would at least have caused the addressee to regard modification to any of [the] prior art proposals with considerable reserve if not scepticism."
  While it is not possible to combine the disclosure of a given prior document with other matter for the purposes of proving anticipation, it is permitted to combine prior art (documents, uses and common general knowledge) in order to prove that an invention is lacking an inventive step. However, to establish that a combination of teachings from the prior art shows an invention to be obvious, it must itself be obvious to combine the prior art. The leading authority for this is *Technograph Printed Circuits v Mills & Rockley* [1972] RPC 346.

8.4   Sufficiency

Section 72(1)(c) of the Patents Act states that a patent may be revoked if the specification of the patent does not disclose the invention clearly enough and completely enough for it to be performed by a person skilled in the art. Thus, as for the enablement requirement of anticipation described above, the ordinary skilled person would have been able to use the disclosure of the patent to perform the invention without undue effort or burden. For this purpose, the skilled person is assumed to be willing to make trial-and-error experiments to get the invention to work.

There is another form of insufficiency attack that has been the subject of recent evaluation by the courts. This is so-called *Biogen* insufficiency, arising from the House of Lords decision of the same name. The *Biogen* principle states that the scope of protection of a claim should be commensurate with the technical contribution of the patent. A claim that claims all processes for making a product, but does not disclose all such processes, is *Biogen* insufficient. The recent Court of Appeal decision in *H Lundbeck A/S v Generics (UK) Limited & Ors* [2008] EWCA Civ 311 has clarified that this principle extends only to process claims or product-by-process claims, not product claims. However, this issue is currently subject to appeal to the House of Lords.

9. Exclusions from patentability

There are a number of categories of invention that are excluded from patentability by the Patents Act. These exclusions are provided for by Section 1(2)(a) (reflecting article 52(2) of the EPC), which excludes: discoveries, scientific theories or mathematical methods; aesthetic creations; schemes, rules or methods for performing a mental act, playing a game or doing business, or a program for a computer; and, the presentation of information. Furthermore, Section 4(A) (introduced by EPC 2000) excludes from patentability methods of treatment of the human or animal body by surgery or therapy and methods of diagnosis performed on the human or animal body. The areas of computer software and biotech inventions have been the subject of recent decisions on how these exclusions apply, as set out next.

9.1 Software

Under Section 1(2)(c) of the Patents Act, programs for computers are expressly excluded from patentability. However, there have been numerous English cases in recent years where patents for computer-related inventions have been considered. The leading authority here is the decision of the Court of Appeal in the joint appeal *Aerotel Ltd v Telco Holdings Ltd and Macrossan's Patent Application* [2007] 1 All ER 225, in which the following four-step test was formulated for determining whether such inventions are patentable:

- properly construe the patent claim;
- identify the actual contribution;
- ask whether the contribution falls solely within the excluded subject matter; and
- check whether the actual or alleged contribution is actually technical in nature.

This approach has been the subject of much discussion. In particular it has been observed that the fourth step is unnecessary. More broadly the four-step test has been criticised as incompatible with the approach adopted by the Technical Boards of Appeal (TBA) – see *Duns Licensing Associates, LP*, Board 3.5.01, T-0154/04, November 15 2006. However, in *Symbian Limited v Comptroller General of Patents* [2008] EWCA Civ 1066, concerning the proper application of the exclusion of computer programs from patentable subject matter, the Court of Appeal confirmed that the four-step test formulated in *Aerotel* was the appropriate test to apply. However, this decision seems to suggest that the third and fourth steps of the test can now be combined, in effect, to ask whether the invention makes a technical contribution to the state of the art. However, on October 22 2008 a number of questions concerning the exclusion of computer programmes as such were referred to the Enlarged Board of Appeal in G3/08. It therefore remains to be seen how UK law will be adapted in response to the decision of the Enlarged Board.

In the meantime it seems that the UK courts will consider the claims of computer-related inventions as a matter of substance rather than form. If the contribution falls solely within excluded subject matter, then it will be excluded. However, if there is a relevant 'technical contribution', the patent will escape exclusion.

9.2 Biotech inventions

Articles 52(1) and 57 of the EPC, corresponding to Sections 1(1)(c) and 4 of the Patents Act, provide that, in order for an invention to be patentable, it has to be capable of industrial application. Furthermore, Section 1(2)(a) of the Patents Act provides that "a discovery ... as such" is not an invention; as Whitford J states in *Genentech* [1987] RPC 553: "It is trite law that you cannot patent a discovery ... "

More specifically, the Biotechnology Directive (98/44/EC) (implemented in Section 76 and Schedule 2 of the Patents Act) states at article 5(1) that elements of the human body, including the sequence or partial sequence of a human gene, are not patentable. However, article 5(2) goes on to say that an element isolated from the human body or otherwise produced by means of a technical process, including the sequence or partial sequence of a gene, may constitute a patentable invention, even if the structure of that element is identical to that of a natural element. So the discovery objection may be overcome by isolation of a gene sequence or protein, for example. However, this still leaves the issue of whether there is industrial application.

Article 5(3) of the Biotechnology Directive states that the industrial application of the sequence or partial sequence of a gene must be disclosed in the patent application as filed if it is to be patentable. However, there is very little case law in the United Kingdom interpreting this requirement. The Court of Appeal briefly considered the point in *Chiron Corporation v Murex and others* [1996] RPC 535 finding that a claim for a virus, which also included polypeptides that might have no relation to the virus, was invalid for lack of industrial applicability. In the recent case of *Eli Lilly and Company v Human Genome Sciences, Inc* [2008] EWCH 1903 (Pat) Kitchin J thoroughly reviewed the EPO authorities on this subject and also looked for guidance to US decisions. Kitchin J stated that simply identifying a protein was not necessarily sufficient to confer industrial application upon it. He also observed that use of the invention as a tool to carry out further research into its activities could not constitute a relevant industrial application.

In other words, speculative uses without accompanying biological data are not enough to warrant patent protection and a monopoly on a gene or protein sequence.

Therefore, in most cases, where the identification of the protein or gene does not immediately suggest an application, further research will be required to identify the function and use of the substance before a patent application should be filed.

9.3 Dosage regime patents

Until the case of *Actavis UK Limited v Merck & Co Inc* [2008] EWCA Civ 444, it had been widely thought that a new dosage regime could not form the basis of a novel and patentable Swiss-type claim in the United Kingdom unless it involved a novel medical indication. This was on the basis that such claims were excluded methods of treatment. However, in *Actavis* the Court of Appeal has now clarified that new dosage regimes and forms of administration of a drug can be novel and are not *per se* unpatentable as methods of treatment. This decision therefore brings the approach of the UK courts into line with the EPO on this important form of claim in the pharmaceutical industry's patent portfolio.

10. Available remedies

The United Kingdom already provides patent holders with a comprehensive set of remedies and was on the whole already in line with Enforcement Directive 2004/48/EC prior to its entry into force in the United Kingdom on April 29 2006. However, the directive does add to existing rights and remedies. The key remedies which may be sought in infringement proceedings before the courts are:

- final injunction and post-expiry injunction;
- an order for delivery up or destruction of infringing goods;
- damages or an account of profits; and/or
- declaratory relief.

10.1 Final injunction and post-expiry injunction

Interim injunctions are described above. Upon a final finding of patent infringement at trial, a perpetual (final) injunction may be ordered against the infringer by the court prohibiting the continuance or repeat of specified infringing acts. As with interim injunctions, this is an equitable remedy and it is at the court's discretion whether or not to grant.

An injunction will normally continue until the right protected ceases to exist. In the case of a patent, this will be at the expiry date of the patent or supplementary protection certificate. However, in the case of *Dyson Appliances Ltd v Hoover Ltd* [2001] RPC 26, the court granted an injunction to prevent the defendant from selling its infringing product for 12 months after the date of expiry of the infringed patent. This post-expiry injunction was granted in this case to provide compensation for the claimant in circumstances where infringing activity took place before the expiry of the patent, such as creating and testing prototypes, submitting samples to relevant regulators and offering the product for sale. These activities had enabled the infringer to ready themselves to compete with the claimant as soon as the infringed patent expired (ie ahead of when they would have been able to market, but for the infringing activities).

10.2   Delivery up or destruction of goods

As an accompanying remedy to injunction, the court has the discretion to order that infringing goods be delivered up, so that they may be destroyed by the claimant, or alternatively that the defendant destroys the infringing goods.

10.3   Damages or account of profits

Typically, there are separate hearings for liability and quantum. The hearing on liability (ie has the patent has been infringed?) takes place first. If the court finds that there has been infringement, then the parties will try to agree an amount of damages that should be paid to the patentee. If no agreement is reached, there will be a separate hearing on quantum to ascertain the amount to be paid to the patentee. In order to determine quantum, the patentee who is successful on liability must elect to claim *either* damages *or* an account of profits, depending on whichever is considered to be larger. Disclosure is available to assist a claimant in deciding which of these remedies to claim.

Damages aim to put the patentee in the position he would have been in but for the infringement, provided that such losses are:

- reasonably foreseeable;
- caused by the infringement; and
- not excluded from recovery by public policy.

Calculating the profits lost by the patentee due to the infringing activity can be difficult and often requires evidence from accountants. There are two methods which are frequently used:

- calculating the lost sales, for example, if the patentee's product is on the market; or
- calculating the licence fee the infringer would have paid the patentee to obtain a licence, for example, if there is evidence that a licence may have been available.

Where the patentee is not able to prove loss of sales and does not grant licences, the court may assess the damages upon a reasonable royalty basis. Exemplary damages are not available as a general rule to punish an infringer for wrongful conduct, although there are narrow exceptions. The court may also award aggravated damages if there has been injury to the claimant's reputation or feelings, where the claimant would instead only be entitled to nominal damages.

An account of profits entitles the patentee to the profits made by the infringer as a result of its infringing activities. In reality, this option may be unsatisfactory if insufficient evidence can be obtained of the infringer's profits, or indeed if those profits are small.

10.4   Declaratory relief

The court has a specific power to grant a declaration of non-infringement under Section 71 of the Patents Act. In order to avail itself of this provision, the applicant must first apply in writing to the patent proprietor for a written acknowledgement of the declaration claimed, providing the proprietor with full details of the non-infringing product or process. Only if the patent proprietor refuses or fails to give any such acknowledgement in a reasonable period of time can the applicant make an application to the court.

Under Section 65 of the Patents Act, where a patent is found valid or partially valid by the court or the IPO, the contested validity of that patent may be given certification. If so, the patentee will be entitled to indemnity costs if the validity of that patent is subsequently challenged and again upheld.

See also Section 12 below.

10.5 Effect of appeal on enforcement?

If a patentee loses at trial, the alleged infringer will ask for an order declaring that the relevant claims of the patent are invalid. However, this order will be suspended if the patentee then appeals. This means that the patent will remain in force pending the outcome of the appeal. Consequently, the patentee could bring infringement proceedings against an alleged infringer who launches its product in the United Kingdom after trial (assuming an infringement action did not already feature at trial). However, it is likely that these proceedings would be stayed pending the outcome of the appeal. It is also unlikely that an interim injunction would be granted if the patent has been held invalid at first instance.

The situation would be different if the Court of Appeal decides that the patent is valid. In those circumstances, if an injunction has not already been obtained at trial, it is possible that the patentee would be able to obtain a preliminary injunction, although this would depend on the facts as they existed at the relevant time (eg whether the alleged infringer had launched or was about to, the state of the market and so on).

11. Costs

The general rule is that the losing party should pay the other side's reasonable and proportionate costs. However, in reality, 100% recovery of costs by the winning party is unlikely. Typically, the winning party can expect to recover 50% to 60% of its legal costs. These costs will include fees for the solicitors (who usually charge an hourly rate) and the barristers (charging an hourly rate for initial work and then a fixed fee for the trial). QCs generally command higher fees than Junior Counsel. Other costs to factor in are court costs, experts' costs and disbursements.

Costs will therefore vary greatly according to the complexity and size of the case, the number of experts used, the value of the patent in question to the patentee, the quality of legal advisers and counsel, the number of issues in the case (ie just infringement or validity also), the number of counterparties and the length of the trial. As an example, the solicitors' and barristers' costs alone for bringing an infringement claim to trial can easily exceed £1 million for a complex case involving a counterclaim for revocation.

12.   Hot topic

12.1   General declaratory relief

In addition to Section 71 of the Patents Act 1977, discussed above, the court also has a general power to grant binding declaratory relief under Civil Procedure Rule 40.20. Recently, litigants in the English courts have devised imaginative ways to use this power to their advantage. For example, the Court of Appeal in *Nokia Corp v Interdigital Technology Corp (Application to Set Aside)* [2007] FSR 27 held that a party may seek a declaration that a competitor's patent is not essential for compliance with certain telecommunications standards. In another case, *Arrow Generics Ltd v Merck & Co Inc* [2007] FSR 39, the High Court refused to strike out an application for a declaration that the manufacture of a certain pharmaceutical at a specified date was obvious. The purpose of such a declaration was to ensure that the products could not be legitimately caught by pending divisional applications once they were granted.

In *Arrow Generics* it was stated that whether or not to grant a negative declaration was a matter of the court's discretion, rather than its jurisdiction. The use of negative declarations should be rejected where it serves no useful purpose, or where the underlying issue had not been sufficiently clearly defined to make it properly justifiable. However, in this case, the declarations had a valuable commercial purpose and a clearly defined purpose had been raised, which was susceptible to determination. These two decisions therefore raise the interesting possibility that further declarations may be sought in future for a whole range of specific purposes outside the scope of traditional non-infringement declarations, providing they have a clearly defined purpose.

United States

James F Haley
William J McCabe
Ropes & Gray LLP

In the United States, patent rights are often one of the most valuable assets a company possesses. Consequently, US patent litigation is one of the most complex and hard-fought areas of US civil litigation. Patent cases are often driven by the goal of securing not only lost-profits damages, but also an injunction that can cripple a competitor.

US patent litigation is nearly the exclusive domain of the federal courts. The lawyers who handle these cases should be experts in both patent law and complex civil litigation. Although only a small percentage of the patent cases filed ever go to trial, it is important to prepare your case with an eye toward trial where the stakes are highest and the outcome can be driven by the groundwork and strategic decisions made in the months or years leading up to trial. In the past several years, certain federal district courts have become favourite venues for patent actions. These courts tend to have a relatively short time to trial, which is perceived as an advantage by the patent owner. The liberal venue and personal jurisdiction requirements in the United States have allowed patent owners a virtually free hand to select where to file a case.

While it may appear that everything favours a patent owner, this is not the case. In addition to the traditional defences of non-infringement and invalidity based on the prior art or a lack of disclosure, a US patent can be held invalid if the patentee failed to disclose in his patent application his best mode of making or using his invention. Moreover, an otherwise valid patent can be held unenforceable if the patent applicant breached his duty of candour during prosecution. Further, patent validity and priority of invention can be challenged in the US Patent & Trademark Office (PTO) through re-examination or interference proceedings. Finally, recent US Supreme Court decisions have favoured those challenging a patent.

1. Lawyers

The United States district courts have original jurisdiction over any action relating to a patent (28 USC § 1338(a)). This jurisdiction is exclusive even as to the courts of the states. In addition to the district courts, the US International Trade Commission (ITC) has authority to enjoin the importation of goods that "infringe a valid and enforceable United States patent" (19 USC § 1337(a)(1)(B)(i)). An infringement action in a district court and an enforcement proceeding in the ITC are similar in that both can adjudicate issues of patent infringement and validity, and both can issue an injunction to prevent future infringement. The district courts, unlike the

ITC, have the authority to award damages for infringement and to address infringement that arises as a result of goods manufactured, offered for sale, or used in the United States as well as those imported into the United States. A patentee may bring an infringement action in the district court regardless of whether it is practising the patented invention. In contrast, in the ITC, a patentee must establish that an industry "relating to the articles protected by the patent ... exists or is in the process of being established" in the United States (19 USC § 1337(a)(2)).

Selection of trial counsel for a patent action is a critical decision that will shape many aspects of a case. Trial counsel should not only be expert in the substantive patent law but also have extensive experience litigating all phases of complex patent matters, from discovery through to trial. Ideally, the trial team will include one or more attorneys with technical training or experience in the subject matter of the patent in suit. The team should also be experienced in efficiently handling the large volume of documents (both paper and electronic) that are now commonplace in US patent litigation. You should not limit your selection of trial counsel to those attorneys who are located geographically close to the courthouse. Instead, you should seek out the best patent trial attorneys to handle your case regardless of where their office is located.

In the United States, attorneys are admitted on a state-by-state basis to practise. Each state has its own bar examination and requirements for admission. An attorney licensed to practise by one state may apply for admittance to practise before the federal courts of that state. There are no additional exams or qualifications. For example, an attorney licensed to practise before the bar of the State of New York may then be admitted to practise before the federal courts within New York State. Further, there is no requirement that an attorney handling a patent case in a district court be registered to practise before the US Patent & Trademark Office as a patent attorney. This credential is only required for practice in matters pending before the US Patent & Trademark Office (eg an interference proceeding, a re-examination, or the filing and prosecution of a patent application).

When an attorney licensed in one state has a matter pending in another state, he must affiliate himself with an attorney who is licensed to practise in that state (unless he is both licensed and has an office in that state). For example, a New York attorney who has a case across the Hudson River in the US District Court for the District of New Jersey must affiliate himself with New Jersey local counsel in order to appear in New Jersey District Court. The New York attorney can be admitted to practise, with local counsel, before the US District Court for the District of New Jersey for a particular case by filing a motion to appear *pro hac vice* (for the purpose of the case) in that court. In essence, the New Jersey local counsel vouches for the New York attorney and is responsible for the New York attorney's conduct. Generally, most district courts will grant a *pro hac vice* motion and treat it as a formality. You should, however, be sure that you are properly admitted before doing anything in the case that requires admittance. For example, some district courts require attorneys to be admitted in their court before taking or defending depositions, even though such activities take place outside of the court and, perhaps, outside of the district. The important thing is to be sure to comply with all of the written and *unwritten* rules of the court in which your case is pending.

You should work closely with local counsel and keep him apprised of your strategy and the motions you intend to file. After all, his name appears on motions and supporting papers. While some local counsel will defer to out-of-town counsel, others will want to be intimately involved in everything that is filed. Neither course can be said to be ideal. The key is to understand the level of involvement your local counsel expects and plan accordingly.

Although retaining local counsel adds a layer of complexity and expense to a case, there is a lot that can be learned from an experienced local counsel that can be of tremendous value in shaping your case. You should carefully choose a local counsel who practises before the court where your matter is pending. It is not critical for local counsel to have extensive (or even any) experience with patent litigation. Rather, good local counsel should have experience appearing before the court, and particularly before the judge who is handling the case, and experience in civil matters, preferably complex litigation. Good local counsel will be able to provide invaluable advice based on his knowledge of the likes and dislikes of the judges of the district and will understand the local rules of the court. Depending on the locale, good local counsel can also help at trial by adding a local flavour to what may be an obviously out-of-town client and, perhaps, your out-of-town accent. Some courts will expect or require local counsel to attend trial, while others will not. If you are in a district or before a judge that expects such participation, plan your trial strategy so that local counsel has the opportunity to participate – for example, examining or cross-examining one or more witnesses. Not every witness in a patent litigation is a technical witness. If you choose your local counsel carefully, the fact that he is not a patent practitioner should not prevent him from participating at trial and enhancing your presentation of evidence.

Finally, it should be noted that there is no required separation of the attorneys who prepare the case and the attorneys who try the case. Because of the extensive amount of pre-trial discovery (both document production and deposition testimony), it would be very difficult (and costly) to transfer all of the pre-trial knowledge of the case to a separate trial team. Generally, the same team that handles the pre-trial preparation of the case also handles the trial.

2. The court system

The United States district courts are trial-level courts. Each of the 50 states has at least one Federal District Court. In addition, the US territories, for example Puerto Rico and the District of Columbia (Washington, DC), each has a federal district court. The larger states have multiple federal districts. For example, New York State is divided geographically into four federal judicial districts: the Southern District of New York (SDNY), the Eastern District of New York (EDNY), the Northern District of New York (NDNY), and the Western District of New York (WDNY). Each of these districts may have courthouses in multiple cities. Further, some districts are divided into divisions. For example, the Eastern District of Texas is composed of the Tyler, Beaumont, Sherman, Marshall, Texarkana and Lufkin divisions. In total, there are 94 federal judicial districts. Each district has one or more federal judges. There are a total of 678 district court judgeships authorised in the United States. The 94 district courts are organised into 12 regional 'circuits' (the District of Columbia and the First to the Eleventh Circuits) and one geographically national circuit with limited subject matter jurisdiction (the Federal Circuit). These circuit courts hear appeals of judgments from the district courts.

The US district courts are courts of general jurisdiction and, as such, hear all types of cases, both criminal and civil. Many district courts have a heavy criminal docket and those matters often take precedence (from a scheduling standpoint) over civil matters such as patent infringement, because of the Speedy Trial Act, which requires trial in a criminal matter within 70 days of indictment (18 USC § 3161). In addition to patent matters, the US district courts hear any matter involving a 'federal question' (ie an issue arising out of federal law). This could include matters as diverse as civil rights, personal injury, fraud, antitrust, labour, securities, social security, tax or environmental issues. In addition, these courts hear any matter that raises a federal constitutional issue, for example freedom of speech or equal protection. Further, the US district courts will hear matters involving state law, if there is a dispute between citizens of different states or if the dispute forms a part of a federal case between the parties. For example, a citizen of New York may sue a citizen of New Jersey in New Jersey State Court for breach of contract (a state law issue), or he may bring the action in the US District Court for the District of New Jersey (provided the amount in issue is at least $75,000) (28 USC § 1332). In such a case, the New Jersey Federal Court may apply New York, New Jersey, or some other state's law as is appropriate to the dispute in issue. This 'diversity' jurisdiction goes back to the early days of the United States, when a citizen of one state did not believe that he would get a fair hearing in another state's court.

District court judges, like all federal judges, are appointed by the President, confirmed by the Senate, and enjoy life tenure. As a result, the federal judiciary has a certain level of independence as their decisions are not influenced by whether they will be reappointed or re-elected. Like the scope of the matters they handle, the background experiences of federal judges are diverse. Some have worked as criminal prosecutors, some as state court judges, others in private practice, and others have held various government or academic positions. It is unusual for a district court judge to have any background in patents or patent litigation. They do, however, have tremendous experience in handling complex matters and trials, and in assessing the credibility of witnesses.

In addition to district court judges, there are magistrate judges. Magistrate judges are appointed by the judge(s) of the district court and serve for a term of eight years (28 USC § 631). Magistrate judges handle many facets of the cases pending before the district courts. The parties may, if they so desire, have a magistrate judge adjudicate their entire matter. Consent to trial by a magistrate judge may speed up the case. Speed, however, is usually an attractive option to only one party.

More often, magistrate judges will handle pre-trial matters, such as adjudicating discovery disputes. In that role, magistrate judges can be very helpful in keeping the matter on course during discovery. In some districts, the magistrate judge may help facilitate mediation of the dispute. Like the district court judges, magistrate judges rarely have any patent experience beyond on-the-job training. In some district courts, for example Delaware, Massachusetts, California, and Texas, that training may be extensive.

Except in patent matters (and other limited areas), an appeal of a district court judgment is taken to the Court of Appeals for the circuit in which the district court is located. For example, the US Court of Appeals for the Second Circuit would hear all appeals of non-patent judgments from the Southern District of New York. The Second Circuit hears appeals from all district courts in the states of New York, Connecticut and Vermont. By contrast, the US Court of Appeals for the Federal Circuit, based in Washington, DC, hears all appeals involving patents regardless of the location of the district court that tried the case. The Federal Circuit was established in 1982 as part of a reform of the US court system. Over the past 25 years, the Federal Circuit has brought a certain level of uniformity to patent law that was believed lacking when an appeal of a patent matter went to one of the twelve regional circuits.

The Federal Circuit has twelve judges who typically sit in panels of three to hear appeals. When addressing a significant issue of law, the Federal Circuit may sit *en banc*, during which all twelve judges will hear the case. An appeal is based on briefs (opening briefs are limited to 30 pages or 14,000 words) and a 30-minute oral argument (if requested). The Federal Circuit applies its 'own' law to patent issues, but applies the law of the regional circuit, where the case originated, to non-patent and procedural issues. For example, on appeal from the SDNY, the Federal Circuit would apply Federal Circuit law to any patent issues and the law of the Second Circuit to other issues (eg interpretation of a licence agreement).

Generally, the Federal Circuit (like the other circuit courts) will only hear an appeal from a final judgment that disposes of the case at the district court level. Cases are usually heard and an opinion issued by the Federal Circuit within about 12 months of the appeal. The Federal Circuit will, however, on occasion, hear certain issues prior to a final judgment of the trial court on interlocutory appeal. For example, if the district court issues an order waiving attorney–client privilege, the aggrieved party may file a mandamus petition asking the Federal Circuit to hear the issue immediately. The Federal Circuit has discretion whether or not to hear any issue before entry of a final judgment.

In addition to hearing all patent appeals from the district courts, the Federal Circuit hears appeals from the US Patent & Trademark Office's Board of Patent Appeals and Interferences, the US ITC, the US Court of Federal Claims (claims against the United States), the Social Security Administration, the Department of Veterans Affairs, and certain other administrative decisions.

Decisions of the Federal Circuit, like decisions from other circuit courts of appeal, may be reviewed by the Supreme Court of the United States. Generally, however, there is no *right* to appeal to the Supreme Court. Instead, a party must petition the Supreme Court to hear its case. Such petition is rarely granted. Between 2002 and 2005, the Supreme Court had between 9,000 and 10,000 cases on its docket each year. About 80% of these were filed *in forma pauperis* (as a pauper) and were typically filed by prisoners seeking to have the Supreme Court hear their case. Of the remaining 2,000 annual cases, the Supreme Court accepted appeals and heard argument on about 80 to 90 cases. Therefore, with few exceptions – about one or two per year – the Federal Circuit is the court of last resort for a patent matter.

Rather than waiting to be sued for patent infringement, a party may file a declaratory judgment seeking a declaration that an adversely held patent is invalid, not infringed, or unenforceable. A recent Supreme Court decision made it easier for a licensee to bring such an action, even while maintaining the licence in good standing (*MedImmune, Inc. v Genentech, Inc.*, 549 US 118 (2007)). Now, a licensee in good standing (ie one who is paying royalties and complying with all of the other terms of the licence) may challenge a licensed patent while still enjoying immunity from an infringement suit or a potential injunction.

A party who initiates a declaratory judgment action may, like any plaintiff, choose the district where the case will be heard. Bringing a declaratory judgment action in your own home district (if personal jurisdiction requirements over your adversary can be met) may allow you as the alleged infringer to blunt the patentee's appeal to a home-town jury regarding, for example, loss of jobs in the local area due to the alleged infringement. The declaratory judgment plaintiff may also be able to maintain plaintiff status at trial, thereby allowing it to open (present his case first) and close (rebut the patentee's case) at trial.

US courts generally will not defer to foreign courts on the interpretation and enforcement of a US patent. US courts have generally held that because different standards are applied to the issue of validity in the United States, foreign decisions on the issue are of no weight. See *Medtronic, Inc. v Daig Corp.*, 789 F.2d 903 (Fed Cir 1986). The same would be true for questions of patent infringement, which involves construction of the patent claims.

3.   **Procedure and timescale of proceedings**

In the United States, the length of time from filing a complaint to resolution on the merits can vary greatly depending on where the patent action is filed. Some jurisdictions are known as 'rocket dockets' because of the short time (one year to 18 months) between filing the complaint and trial. Such jurisdictions include the Western District of Wisconsin (Madison), the Eastern District of Texas (Marshall and Tyler), and the Eastern District of Virginia (Alexandria). These courts are perceived to be plaintiff friendly. This perception may be due to the fact that the short time to trial may not give a defendant adequate time to develop its defences. The plaintiff presumably has the advantage of preparing his case before filing his complaint and the shorter time to trial is a disadvantage to the defendant. Other jurisdictions are known for taking longer to get to trial.

A patent owner may bring suit in any judicial district where the defendant resides or where the defendant has allegedly committed acts of infringement and has a regular and established place of business (28 USC § 1400). An alien (eg a foreign corporation) may, however, be sued in any district (28 USC § 1391(d)).

In 2007, the most popular jurisdictions for patent cases (based on the number of cases filed) were the Eastern District of Texas (359 cases), the Central District of California (334), the District of New Jersey (186), the Southern District of New York

(111), and the District of Massachusetts (69).[1] With the exception of the Eastern District of Texas, each of these districts is home to numerous companies.

A typical patent litigation can be divided into four phases:

- pleading and initial disclosures;
- discovery;
- *Markman* claim construction and summary judgment; and
- trial.

The case begins with the formal pleadings of a complaint and answer. The defendant may plead a counterclaim (eg assertion of its own patents or other rights) (Fed R Civ P 7). These pleadings tend to be notice, bare-bones papers. Next, each party makes its initial disclosures in which documents and witnesses with relevant knowledge are identified (Fed R Civ P 26(a)).

A court will then generally hold a 'Rule 16' conference to address the overall schedule and the scope of discovery. This is usually the first opportunity for the parties to inform the court of any issues that might justify a pre-trial schedule that differs from the court's usual standing order for the pre-trial activities in a case. For example, if the case involves non-US parties or numerous non-English-language documents, the court should be informed so that extra time can be built into the schedule to allow for travel for depositions or translation of documents (Fed R Civ P 16).

Following initial disclosures and the Rule 16 conference is a discovery period during which document discovery and depositions take place. This discovery period is typically the most time consuming (and accordingly expensive) part of the pre-trial phase of the case. Each party has an obligation to locate and identify documents (and things) relevant to its claims or defences. Each party may serve document requests and interrogatories (written questions) seeking information from the other side. In order to respond to these requests and to develop your case for trial, you must identify, collect and review documents, including electronic documents, from your client (Fed R Civ P 30–34).

As explained in detail below, discovery in US civil litigation is very broad. Attorneys have a duty to carry out a good-faith search for relevant documents and things. Once the documents are located and collected, they must be reviewed for privilege. Careful attention to this privilege review is very important. Any documents withheld on the basis of privilege must be identified to an adversary (Fed R Civ P 26(b)(5)). Some district courts will find a waiver of the attorney–client privilege if privileged documents are produced inadvertently. Production of a single privileged document can result in waiver of privilege for the subject matter to which the document relates. For example, production of a privileged document containing advice regarding patent invalidity or infringement can result in a court order to produce all privileged documents relating to those subject matters. While such a broad waiver rarely happens as a result of inadvertent production, such production

---

will almost guarantee time-consuming motion practice regarding the waiver and its scope (see Fed R Evid 502).

Because much of the information that is collected and produced by the parties is confidential and non-public, most parties will negotiate a protective order that the court will enter. As discussed in some detail below, that protective order will govern the disclosure and use of a party's confidential information. Such an order might limit disclosure of certain information to outside counsel or outside experts who sign an agreement not to disclose the information or use it for purposes other than the litigation.

The Rules of Civil Procedure were recently amended specifically to address discovery of electronically stored information (Fed R Civ P 34). While there is recognition that the volume of information stored in electronic form can be quite significant, there remains an obligation for the producing party to search for and produce relevant information. Importantly, there is an obligation for a party involved in litigation or a party that reasonably expects to be involved in litigation to put a 'litigation hold' on any document destruction that might normally occur. This practice helps to ensure that relevant documents (eg e-mails) that might normally be erased after some set period of time are preserved. A litigation hold should be clearly communicated to everyone in the company who might be involved in any issue arising in the case. Importantly, a litigation hold should apply to both electronic and non-electronic documents. Recently, courts have ordered sanctions where electronic documents were either not properly preserved or produced (see *Qualcomm Inc. v Broadcom Corp.*, No. 05-CV-1958 (SD Cal August 6, 2007)).

In addition to document discovery, a party may take up to ten depositions (oral examination of witnesses outside of court) without leave of court (Fed R Civ P 30(a)(2)(A)(i)). A party may depose another party's witnesses or may take the deposition of a non-party. The deposition of a non-party is typically accomplished by subpoena (Fed R Civ P 45). This pre-trial testimony may later be used at trial if the witness is not available or cannot be compelled to attend. Relevant portions of the pre-trial deposition may be read into the record or, if the deposition was videotaped, the tape may be played for the judge and jury. For witnesses who are available and testify at trial, deposition transcripts or videotapes may be used for impeachment.

In addition to fact witnesses, most parties will present expert witnesses in support of their case. Typically, a party will have an expert on the issues of patent validity and infringement, and an expert on the issue of damages. During the pre-trial stage of the case, each party is required to provide "a complete statement of all opinions the [expert] will express and the basis and reasons for them" (Fed R Civ P 26(a)(2)(B)(i)). Some judges are more strict than others in limiting an expert's trial testimony specifically and exactly to the opinions expressed in the report. The best practice is to prepare expert reports with the expectation that the judge will limit the experts' testimony to the opinions expressed in the report.

While the rules allow for court-appointed experts (Fed R Evid 706), such appointments are rare. Most courts prefer to allow the parties to select their own experts and will weigh the expert testimony as needed. Courts will, on occasion, appoint a special master to hear certain aspects of a case and make recommendations to the court.

In the United States, the interpretation of the meaning and scope of a patent claim is a question of law reserved for the judge. The process of construing the claims is known as a *Markman* proceeding after *Markman v Westview Instruments*, 517 US 370 (1996). Claim construction is a critical part of any patent case because the judge's construction of key claim terms often results in summary judgment of invalidity or non-infringement, thereby ending the case. Each party will typically submit written arguments based on the intrinsic evidence (eg the patent and its file history) and, on occasion, extrinsic evidence (eg dictionaries or treatises) in which they argue how the claims should be construed. These submissions may be accompanied by an expert declaration explaining the meaning of certain terms as used in the art at the time the application that issued as the patent was filed. *Phillips v AWH Corp*, 415 F3d 1303 (Fed Cir 2005) (*en banc*). In some courts (eg the Northern District of California) the procedures for claim construction are set out in the local rules.

It is not unusual for a court to hold a *Markman* hearing during which the court will hear oral argument on the claim terms. Some courts do so in the context of a motion for summary judgment. *Markman* hearings can last from an hour to a day depending on the number and complexity of the terms at issues. On occasion, a court may wish to hear testimony from the experts. After the hearing, the court will typically issue a *Markman* order setting forth the court's interpretation of the patent claim terms in dispute. This construction controls the evidence and issues for the rest of the case.

4.   Availability of pre-action evidence gathering/disclosure

Generally, discovery is not available until a complaint is filed. There are, however, exceptions. For example, a party may petition the court for an order to preserve testimony before an action is initiated, provided the party can establish that the testimony is relevant to an action that will be brought and that the party seeking the testimony cannot presently initiate the action (Fed R Civ P 27). Testimony preserved in this manner may be used like any other deposition taken in the case.

It is not unusual that discovery taken in a US action can shed light on or otherwise impact a matter pending in another non-US jurisdiction. For example, knowledge of an adversary's weaknesses obtained through discovery in a US action may drive settlement negotiations or resolution of matters pending in another non-US jurisdiction. Use of discovery in another matter, however, may be restricted by a protective order entered in the US case. You must be careful not to use US discovery in a way that would violate a protective order.

Moreover, discovery may also be available in US courts in aid of non-US legal proceedings. The district court may order a person to give his testimony or statement, or to produce documents or things for use in a foreign proceeding. The order may be made pursuant to a letter rogatory, a request made by a foreign or international tribunal, or upon application by an interested person (see 28 USC § 1782).

5. Availability of interim relief (especially injunctions)

In the United States, a patentee may seek interim relief in the form a preliminary injunction (Fed R Civ P 65). The purpose of a preliminary injunction is to preserve the *status quo* pending resolution of the issues of patent infringement and validity on the merits.

To obtain a preliminary injunction, the patentee must establish a substantial likelihood of success on the merits. That is, the patentee must show that he will prevail on the issue of infringement and that he will suffer irreparable harm absent an injunction. The accused infringer is given the opportunity to raise a substantial question regarding the validity or infringement of the patent, or to show that there is no irreparable harm.

The irreparable harm prong of the preliminary injunction test usually drives the analysis. For example, a patentee who has licensed or is willing to license his patent can usually be made whole with money damages. Accordingly, a court is unlikely to find irreparable harm in those situations. Similarly, a patentee who was aware of the infringement but waited a significant period of time to file suit or to request the preliminary injunction is unlikely to be found to be irreparably harmed. A typical case in which irreparable harm is found is one in which a competitor announces a new product (eg at a trade show) and the patentee moves quickly to stop the marketing and sale before the accused infringer enters the market. The converse of that would be when a patent is granted and the patentee quickly sues his competitor for infringement of the newly granted patent.

Because a patent is presumed valid, the burden of establishing invalidity falls on the accused infringer, who must present clear and convincing evidence of invalidity (ie evidence that it is highly probable). This burden can be very difficult to meet at trial even after full discovery from the patentee and an opportunity to search for prior art. At the preliminary injunction stage, the accused infringer bears the burden of raising a substantial question of validity within weeks of learning of a patent and without full discovery or the opportunity to search fully for prior art. Thus, meeting this burden to prevent a preliminary injunction is difficult.

Although a request for a preliminary injunction is not the norm, the grant of such an injunction can dramatically change the posture of a case. The accused infringer is effectively out of the market pending trial, which may not occur for a year or more. On the other hand, denial of a preliminary injunction can have a severe impact on the patentee. In denying the requested injunction, the court may decide that the scope of the patent claims is more narrow than the patentee advocates and, hence, not likely to be infringed, or that the claims are unlikely to be valid. Such a ruling can have a major influence on any settlement talks between the parties.

6. Disclosure

The broad scope of disclosure is one of the hallmarks of litigation in the United States. At the start of a case, each party is required to provide initial disclosures to the other that include "the name and, if known, the address and telephone number of each individual likely to have discoverable information – along with the subjects of that information – that the disclosing party may use to support its claims or defenses". In addition, each party must provide "a copy – or a description by category and location – of all documents, electronically stored information, and tangible things that the disclosing party has in its possession, custody, or control and may use to support its claims or defenses" (Fed R Civ P 26(a)(1)(A)(i), (ii)).

Initial disclosures are followed by a period of discovery during which a party may seek documents from another party or a non-party, interrogatories may be served, and depositions taken. Discovery in the United States is very broad. A party "may obtain discovery regarding any nonprivileged matter that is relevant to any party's claim or defense – including the existence, description, nature, custody, condition, and location of any documents or other tangible things and the identity and location of persons who know of any discoverable matter" (Fed R Civ P 26(b)(1)). 'Relevant information' here is not limited to information that is admissible at trial. Rather, it extends to information "if the discovery appears reasonably calculated to lead to the discovery of admissible evidence" (Fed R Civ P 26(b)(1)).

Although discovery is broad, there are some limitations. For example, without leave of court, a party is limited to ten depositions of seven hours on-the-record each and no more than 25 written interrogatories. These limitations may be expanded by agreement of the parties, with the court's consent depending on the scope of the case. For example, in a case involving multiple patents with multiple inventors the allotted ten depositions could be used up with only the depositions of the inventors.

In addition to depositions of individuals, a party may take the deposition of a company or organisation. The party taking such a deposition "must describe with reasonable particularity the matters for examination". The party representing the company must then designate a witness to testify on behalf of the organisation. The witness "must testify about information known or reasonably available to the organization" (Fed R Civ P 30(b)(6)). This procedure can be a very effective method for obtaining information from a company on very specific topics via deposition, because it requires the organisation being examined to put forth a witness knowledgeable on the identified topics.

Protection of confidential information by use of a protective order is commonplace in US civil litigation. If the parties cannot reach agreement on the terms of a protective order, the court will resolve the issues. Generally, it is best if the parties reach agreement on as many issues as possible and only present to the court those issues where agreement truly cannot be reached.

In most cases, the parties will negotiate a protective order that may treat various categories of documents differently. For example, technical information may be treated differently from competitive sales and marketing data. Outside experts and outside counsel are usually given access to all of the confidential information. Outside experts will typically sign an undertaking agreeing to be bound by the provisions of the protective order. In-house counsel may be given limited access to certain types of information. Although no-one wants his competitors to have access to confidential information, allowing in-house counsel some level of access may help with settlement negotiations because they will be better informed of what is happening in the case. This limited access may be further tempered if the in-house (or even outside) counsel is handling patent prosecution. Competitors may want such counsel to choose between access to a competitor's confidential information or patent prosecution. Ethical walls may be set up to restrict the flow of confidential information to persons who are prosecuting patents for a competitor.

It is important not to use discovery from an adversary in a way that is not in accordance with the protective order. For example, documents that are designated as confidential must only be used with witnesses who are entitled to access such documents, or whose names appear on the documents themselves as author or recipient. If confidential documents or testimony are used in support of a motion, they must generally be filed under seal to protect them from becoming public. At trial, however, the use of confidential information is generally not so restricted. Because the courts are generally open to the public, especially during trial, confidential information either from testimony or documents may become part of the public record (Fed R Civ P 77(b); "[e]very trial on the merits must be conducted in open court"). In the rare instance where a trade secret (eg secret formula or production process) may become public during trial, the court should be made aware of the issue ahead of time and efforts should be made to avoid explicit disclosure.

7.   Evidence

Trial in the United States centres around the testimony of fact and expert witnesses. Such testimony may be supplemented or aided by documentary or physical evidence. For example, an inventor may testify about the origin of his invention while using his notebook. He may show physical samples of his invention to help explain the technology. There are, of course, limits to what may be admitted into evidence. For example, certain types of hearsay – out of court statements offered for their truth – are inadmissible and may be objected to (Fed R Evid 801). Other evidence may be excluded because it is unduly prejudicial or likely to confuse the jury (see Fed R Evid 403).

In cases where a jury is not hearing the evidence, some courts will require a witness's direct testimony to be submitted by written statement in advance of trial. The witness would then be called and cross-examined on the stand at trial. Presenting direct testimony by written statement is often used in ITC proceedings.

As stated earlier, in a typical patent case each party will have two or more experts – one or more on the patent liability issues (infringement and validity) and one with respect to damages issues. Under the rules, a party must fully disclose the experts' opinions during the discovery phase. The opponent then has an opportunity to examine the expert at deposition. Parties hire and compensate their own experts. The qualifications and opinions of an expert are subject to challenge at trial (*Daubert v Merrell Dow Pharms. Inc.*, 509 US 579 (1993)).

Often, an expert may be asked to perform tests or experiments. For example, an expert may need to test the accused product in order to form an opinion as to whether the product meets the patent claim limitations and thus infringes. Or an expert may need to run an experiment to test the prior art. These tests or experiments are fully discoverable by an adversary if they are carried out by a testifying expert (Fed R Civ P 26(b)(4)(A)) (ie an expert who has been designated to testify at trial). If trial counsel is trying to understand the patent claim limitations as applied to the accused product or is trying to determine whether the prior art performs in a certain manner, he would be well advised to have a non-testifying expert conduct a dry run before asking a testifying expert to do so. Tests or experiments by a non-testifying expert done at the request of counsel may be protected by the work product doctrine and are not discoverable by your adversary (Fed R Civ P 26(b)(4)(B)).

Ultimately, counsel may want to rely on tests in the party's evidence. The results of those tests should be presented in an expert report during the discovery phase of the case. There may, however, be some tests that are simple enough and compelling enough to run again in court. When faced with such a situation, you should consider videotaping the test beforehand as an evidentiary back-up. While this gives your adversary a look at your tests that he might not otherwise have, it will outweigh the downside of the test failing during trial. An additional advantage of a videotaped test is that your expert can focus on narrating the test and not on running the test during trial. Videotaping would apply, obviously, to tests that cannot be run in court because of physical or other limitations.

8.   Law

Subject matter that is eligible for patent protection in the United States is quite broad (see 35 USC § 101). The US Supreme Court commented that "Congress intended statutory subject matter to 'include anything under the sun that is made by man'" (*Diamond v Diehr*, 450 US 175, 182 (1981)). Patent protection has been granted for software, biotechnology, and medical techniques including living non-human organisms and methods of treatment. The limits of the scope of protection in the context of business method patents was recently addressed by an *en banc* panel of the Federal Circuit in *In re Bilski*. There, the Federal Circuit "reaffirm[ed] that the machine-or-transformation test, properly applied, is the governing test for determining patent eligibility of a process under § 101" (*In re Bilski*, No. 07-1130, slip op. at 15 (Fed Cir October 30, 2008)).

In the United States, the issues of patent infringement and validity are addressed together. The first step is for the court to construe the meaning and scope of the claims. As described above, this process is known as the *Markman* proceeding. Claims are given the same scope for infringement and validity purposes.

To infringe a claim, a patentee must prove by a preponderance of the evidence (more likely than not) that every element of the properly construed claim is found in the accused product or process. If a single claim element is missing, there is no literal infringement. Infringement may also be found under the doctrine of equivalents (DOE). Under the DOE, infringement may be found where the differences between a claim element and the accused product or process are insubstantial. One test for infringement under DOE is the 'function-way-result' test (*Hilton Davis Chem. Co. v Warner-Jenkinson Co.*, 62 F.3d 1512 (Fed Cir 1995) (*en banc*)). The question is whether the accused device performs substantially the same function in substantially the same way to achieve substantially the same result as compared with the claim element at issue. Significantly, the scope of equivalents permitted under the DOE can be narrowed or eliminated based on the patent's prosecution history. In short, a patentee cannot extend the scope of his claim to cover by equivalents subject matter that was surrendered during prosecution.

The United States, like most of the world, requires that patent claims be both novel and non-obvious. However, unlike most jurisdictions, the United States allows the patentee a one-year grace period to file a United States patent application measured from first public disclosure or offer for sale of the claimed invention. A patentee is permitted to 'swear behind' prior art that arises in the intervening year if he can demonstrate that he invented the claimed invention before the effective date of the prior art. After one year, the patentee's own public disclosures become prior art even as to the patentee (35 USC § 102(b)).

Analysis with respect to novelty is straightforward and requires that a single document or public use embody every element of the patent claim either explicitly or inherently.

Analysis with respect to obviousness is, not surprisingly, more complex. The scope and content of the prior art and the level of a person of ordinary skill must be established. Then, the issue of what would have been obvious to that skilled person must be evaluated and objective evidence of obviousness or non-obviouness (eg long-felt need or commercial success) considered. When combining prior-art references, the Federal Circuit had a long history of requiring a showing of a teaching, suggestion or motivation in the art to make the combination in order to conclude that a claim was obvious. In *KSR Int'l Co. v Teleflex Inc.*, 127 S Ct 1727 (2007), the Supreme Court rejected the Federal Circuit's rigid application of such a test and stated: "Common sense teaches, however, that familiar items may have obvious uses beyond their primary purposes, and in many cases a person of ordinary skill will be able to fit the teachings of multiple patents together like pieces of a puzzle" (127 S Ct at 1742).

The impact of the *KSR* decision is still being felt in the district courts and in the Federal Circuit. Certainly, it is making the case easier of those challenging patent validity on the ground of obviousness.

Unique to US patents is the requirement that a patent specification "set forth the best mode contemplated by the inventor of carrying out his invention" (35 USC § 112)(¶ 1). A patent may be found invalid if a patentee had a best mode and failed to disclose it in his patent specification.

9.   Available remedies

A prevailing patentee in a district court action is entitled to seek both an award of damages and an injunction against future infringement.[2] The Patent Act mandates "damages adequate to compensate for the infringement, but in no event less than a reasonable royalty for the use made of the invention by the infringer, together with interest and costs as fixed by the court" (35 USC § 284). A patentee will, however,

---

2   As explained above, a prevailing patentee in an ITC proceeding is not entitled to money damages, only an exclusion order. Similarly, damages are typically not available for an infringement action brought under 35 USC § 271(e) because there must have been commercial manufacture, use or sale of an approved drug.

James F Haley, William J McCabe

often seek compensation greater than the minimum of a reasonable royalty. A patentee may seek lost profits due to lost sales or price erosion. In either case, the patentee must generally prove by a preponderance of the evidence that the lost sale or price erosion was the direct result of the infringing activity. That is, the patentee must show that, but for the infringement, he would have made the sale, or made the sale at a higher price. An accused infringer will often point to factors, other than the patent, to explain the lost profits. For example, the accused infringer might point to the fact that the market perceived the patentee's goods or services as inferior and purchased the infringer's goods because of their perceived better quality or service. Often, this type of defence can only be developed as a result of discovery of the patentee's own customer records.

With respect to a 'reasonable' royalty, the royalty rate is based on a hypothetical negotiation between the patentee, who is a willing licensor, and the accused infringer, who is a willing licensee, at a point in time immediately before infringement begins. This negotiation takes into account the so-called *Georgia-Pacific* factors,[3] which include what others have paid for such a licence (or a similar licence in the field), the cost of designing around the patent, and the types of non-infringing alternatives that are available.

Pre-suit damages for infringement, however, can be barred if a patentee is making, offering for sale or selling goods covered by the patent in suit and the patentee has failed to mark those goods with the patent number (35 USC § 287). In that case, damages only begin to accrue as of the date the complaint is filed or the date the patentee provided notice of infringement to the alleged infringer, not merely notice of the patent.

10. Costs

US patent infringement has been called the 'sport of kings' because of the cost and unpredictability involved in bringing a case to trial (*The Economist*, 60–61 (March 8, 2003)). A large part of that cost stems from the extensive pre-trial discovery that typically occurs and from the trial itself. Both of these phases demand a large amount of attorney time and attention. In addition to the cost of attorneys (and paralegals), there are significant non-attorney costs during discovery for depositions (court reporters and videographers), experts, prior-art searches, document handling (copying, translation, coding and so on), and travel. Trial also involves significant additional out-of-pocket costs.

Various studies report that it can cost in excess of $4 million to bring a patent case to trial in the United States. The American Intellectual Property Law Association (AIPLA) publishes data regarding the costs associated with different intellectual property matters. AIPLA's 2007 survey reports that for a patent infringement action where there is from $1 million to $25 million at risk, the average cost through to discovery (not including trial) is about $1.6 million, while the average total cost for such a case, including trial, is about $2.5 million. For patent litigation where there is more than $25 million at risk,

---

3   *Georgia-Pacific Corp. v U.S. Plywood, Inc.*, 318 F Supp 1116 (SDNY 1970), *aff'd sub nom. Georgia-Pacific Corp. v U.S. Plywood-Champion Papers, Inc.*, 446 F.2d 295 (2d Cir. 1971).

AIPLA reports that the average cost through to discovery is $3.3 million with an average total cost through to trial of $5.6 million.

As noted, these are average costs for various patent cases. The total cost in any particular case can vary widely (perhaps 1.5× to 2× the average) depending on the number of parties and patents involved, the complexity of the technology, and the scope of the pre-trial discovery.

11. Hot topics

Several aspects of US patent litigation distinguish it from litigation in other jurisdictions. The most stark difference, perhaps, is the availability of trial by jury. In the United States, the jury hears the evidence and resolves disputed issues of fact. Questions of law are reserved for the judge to resolve. Although a jury is requested more often than not in patent cases, it is not (usually) available in an Abbreviated New Drug Application (ANDA) case. An ANDA case provides a process by which a generic drug manufacturer seeks to introduce a previously approved drug to the market by challenging the validity or infringement of the branded company's patents that are listed in the FDA's Orange Book in connection with the approved drug product.

In addition to the district courts, the validity of a patent can be challenged through an interference or re-examination proceeding. These proceedings are conducted in the US Patent & Trademark Office and resolve questions of priority of invention and validity (interference) and validity over certain types of prior art (re-examination).

Two other distinctive aspects of US patent litigation are wilful infringement and inequitable conduct. An infringer who is found wilfully to have infringed can face the prospect of enhanced damages (up to treble the actual damages). Conversely, a patentee who has been found to have breached his duty of candour to the Patent Office can be found to have committed inequitable conduct. A finding of wilfulness or inequitable conduct can lead to an award of attorney's fees.

11.1 Jury trials

In the United States, the right to a trial by jury for civil matters is guaranteed for suits at common law "where the value in controversy shall exceed twenty dollars" (US Const amend VII; Fed R Civ P 38). This guarantee has led to the use of juries in many patent cases, when demanded by a party. The Federal Judicial Center, which compiles and publishes statistics on various cases handled by the federal courts, reports that, from 2001 to 2006, only about 3.5% to 4% of the 2,500 to 3,000 patent cases filed annually went to trial either by a judge or jury. However, of those trials, 65% to 75% were tried before a jury.[4] Therefore, the vast majority of patent cases (96%) were disposed of prior to trial – for example, summary judgment, dismissal or settlement. But, those that do go to trial are typically tried to a jury.

Jurors for cases in the United States are selected from the pool of citizens who reside

---

4   By comparison, the Federal Judicial Center reports that only 1.4% to 2.0% of all civil cases filed in federal court reached trial during that same period. Data from the Federal Judicial Center US District Courts – Civil Cases Terminated, by Nature of Suit and Action Taken, During the 12-Month Period Ending December 31, 2001–2006.

in the geographic area of the district in which the court has jurisdiction. Jurors who are summoned to court may be selected for criminal or civil cases. There are no special qualifications to sit on a jury for a patent matter or any other matter. Accordingly, the educational background of jurors varies from some high school to a PhD. There are few exclusions from jury duty. For example, attorneys may serve on juries.

A court will seat enough jurors (but no more than 12) at the start of a trial to ensure that there will be at least six jurors remaining to deliberate in order to avoid a mis-trial (Fed R Civ P 48). In a federal court, the judge typically conducts *voir dire* of the jurors using questions submitted by counsel. Such questions should be designed to elicit responses that help evaluate whether the juror has any prejudice that might negatively impact the case. The court may then dismiss certain jurors for cause (eg a juror whose relative works for one of the parties). For a two-week trial, a court might seat eight jurors to ensure that at least six will remain through to the end of trial. In such a case, the court may initially select fourteen and allow counsel for each party to strike three jurors each. The remaining eight jurors would then hear the evidence, deliberate, and render a verdict. The jury's verdict typically addresses at least infringement, validity and damages.

Interestingly, given that most cases are tried in front of local juries, commentators have reported that the 27% 'win rate' for non-US companies as patent owners is slightly higher than the overall win rate for US patentees. Similarly, the 78% win rate for foreign companies accused of infringement is slightly higher than the overall win rate of accused US infringers (Paul M Janicke and LiLan Ren, *Who Wins Patent Infringement Cases?*, 34 AIPLA QJ 1, 22 (2006)).

11.2 Abbreviated new drug applications

The 1984 Drug Price Competition and Patent Restoration Act (known as "the Hatch-Waxman Act") provides a process by which a generic version of a previously approved patented drug can be introduced to the market. The Food and Drug Administration (FDA) publishes patent information for approved drugs provided by the branded company in its "Approved Drug Products with Therapeutic Equivalence Evaluations" (known as 'the Orange Book'). A generic drug manufacturer is permitted to submit an abbreviated new drug application (ANDA) which refers to a previously approved drug ('the listed drug') and certifies, *inter alia*, that the active ingredient is the same as in the listed drug (21 USC § 355(j)).

As part of its ANDA, the generic drug manufacturer must submit certain certifications regarding the patents listed in the Orange Book for the drug product of interest. A generic manufacturer will often submit a 'Paragraph IV certification' that the listed patent at issue is either invalid or not infringed by the drug product for which approval is sought (21 USC § 355(j)(2)(A)(vii)(IV)). Such a certification is an act of infringement under (35 USC § 271(e)(2)), and it triggers jurisdiction of the district court to adjudicate the issues of patent validity and infringement.

The first generic drug manufacturer to file a substantially complete application and a Paragraph IV certification (all same-day filers are 'first') is granted a 180-day period of exclusivity if it prevails on the issues of invalidity or non-infringement and is granted tentative approval to market the drug within a time period set by statute

(21 USC § 355). This period of exclusivity can be extremely valuable to a generic drug manufacturer. As the first to enter the market, the generic company needs to compete only with the branded company for market share and profit.

A generic drug manufacturer cannot manufacture or sell a generic version of a patented drug product without approval of the FDA. That approval is not available until adjudication of the patent issues or after a time period set by statute. As a consequence, no damages typically accrue prior to trial in a Paragraph IV certification case. Without damages, there is no entitlement to a jury in such an ANDA case.

11.3   Patent interferences

Interferences are peculiar to the United States and Canada (conflicts) (see 37 CFR § 41.200 and following). They reflect the 'first to invent' versus 'first to file' basis for patent grant in the United States (35 USC § 135). A potential interference should always be considered in any pre-litigation strategy.

Interferences are conducted before the PTO's Board of Patent Appeals and Interferences and principally determine priority of invention between two patent applicants, or between a patentee and a patent applicant, who are claiming the same patentable invention (37 CFR § 41.203). Priority contests between two patentees can only be had in the federal courts (35 USC § 291). Interferences, however, can also address issues, in addition to priority of invention. Among these are patentability of a party's claims over the art and sufficient disclosure (written description, enablement and best mode) of the involved application or patent.

Interferences are best handled by specialised counsel skilled in interference and litigation practice. To act in an interference, counsel must also be registered to practice in the United States Patent & Trademark Office. Interferences are less costly than patent litigation. However, costs in the order of several millions of dollars should be expected in highly contested interferences.

Typically, interferences have two phases. Each is conducted before an Administrative Patent Judge (APJ) of the Board of Appeals and Patent Interferences. Each phase lasts about one year.

In the first phase (preliminary motions), the contours of the interference, as authorised by the APJ, are addressed. The parties set forth the best dates of invention that they believe they can prove: conception (a definite and permanent idea of the complete and operative invention) and reduction to practice (actually making and perhaps testing the invention or filing a patent application describing it). The count, defining the interfering subject matter, is constructed. Each party's claims that are unpatentable over, and thus correspond to, the count and which would be lost on an adverse priority decision are designated. Priority support in earlier patent applications for an embodiment of the count and, thus, the parties' entitlement to priority from those applications are adjudicated. Finally, the patentability of claims that correspond to the count may be determined. These procedures typically involve written motions, expert declarations and cross-examinations. Ultimately, there is oral argument before a three-member panel of the Board and a written decision.

The second phase of an interference is the priority phase. In this phase, the parties present their evidence on priority of invention as to the count adjudicated in the first phase of the interference, and sometimes on other issues, as authorised by the APJ for example, the unpatentability of an opponent's claims that correspond to the count. Here, notebooks and other research records are put into evidence by the declaration or deposition testimony of appropriate witnesses. Again, there are motions setting forth the parties' positions on the issues and cross-examinations of the parties' evidence and witnesses. Finally, there is oral argument and a decision by the Board.

Appeals from interference decisions may be taken either to the Federal Circuit (on the record developed before the Board) or to a Federal District Court (and then subsequently to the Federal Circuit). The latter route of appeal is called a '146 action' and has many of the hallmarks of patent litigation (see 35 USC § 146). Owing to jurisdictional restraints, 146 actions are often tried in the District Court for the District of Columbia.

11.4 Patent re-examination

In the United States, a patent may be challenged in the PTO on certain grounds (unpatentability over patents and printed publications) after grant (see 35 USC § 301 and following). These proceedings are called re-examinations. They are conducted by the Central Re-examination Unit of the PTO before an examiner who was not involved in the prosecution of the application that issued as the patent under re-examination.

Re-examinations are often used in place of, or more typically in conjunction with, patent litigation. They allow either the patentee or a third party to have the PTO decide the patentability or unpatentability of the issued claims in light of printed publications and patents if they present a substantial new question of patentability. Re-examinations are typically faster than patent litigation. The evidentary standard for challenging patent validity is also lower in the PTO – a preponderance rather than clear and convincing evidence is needed. Indeed, some courts stay pending litigation in view of re-examination proceedings (see 35 USC § 318). Decisions in re-examination often also lead to the settlement of the companion litigation or perhaps its dismissal or no litigation at all.

For a re-examination to be initiated, the PTO must first find that there is a substantial new question of patentability as to at least one claim of the patent. Thus, printed publications and patents that the examiner did not consider during initial prosecution are the preferred evidence. However, re-examinations have been initiated on the basis of prior art cited during initial prosecution. Indeed, re-examinations based on prior art that was found not to invalidate the patent in an infringement suit have been initiated.

There are two types of re-examinations, *ex parte* and *inter partes*. *Ex parte* re-examination can be initiated by either the patentee or a third party. A third party's participation in an *ex parte* re-examination is limited to actions before substantive re-examination begins. The third party also has no right to appeal if patentee prevails in the re-examination (the third party may only petition the Commissioner of Patents if re-examination is not declared).

Inter partes re-examinations (only available for patents granted from applications filed on or after November 29, 1999) can be initiated only by third parties. They allow extensive participation by the third party during the substantive re-examination proceedings. However, the third party is estopped in later infringement litigation from relying on printed publications and patents that could have been raised in the *inter partes* re-examination. To date, about 400 *inter partes* re-examinations have been requested.

Re-examinations are much less costly than patent litigation (about 5% to 10% of such cost). The success rate in having a re-examination declared is high (above 90%). And, importantly, the success rate in the context of invalidating (ie cancelling) or at least forcing a patentee to amend one or more of its patent claims in *ex parte* re-examinations is about 75% (claims wholly cancelled in 10% of all re-examinations).[5] In the first 30 *inter partes* re-examinations, 73% of the claims attacked by the requestor have been cancelled or disclaimed.[6]

Appeals from re-examination can be taken only by the patentee in an *ex parte* re-examination, and by either party in an *inter partes* re-examination. They are initially heard by the Board of Patent Appeals and Interferences. Further appeals may be taken to the Federal Circuit in *ex parte* re-examinations filed on or after November 29 1999, and for re-examinations filed before that date an appeal can also be taken to the District Court for the District of Columbia, and then to the Federal Circuit. In all *inter partes* re-examinations, either party may appeal from the Board's decision, or participate in the other's appeal, only to the Federal Circuit (third parties have the right to appeal or to participate in appeals only in *inter partes* re-examination commenced after November 2 2002).

11.5 Wilfulness

Patent infringement is a strict liability tort. There is no need to demonstrate intent to infringe or even knowledge of the patent to prove infringement. However, when an infringer is aware of a patent and behaves in an objectively reckless manner, he may be found to be a wilful infringer (*In re Seagate Tech.* LLC, 497 F.3d 1360 (Fed Cir 2007) (*en banc*)). A finding of wilful infringement under such circumstances can lead to the court enhancing the amount of damages awarded up to treble (35 USC § 284). The court may also use a finding of wilfulness to declare the case exceptional and make an award of attorneys' fees to the patentee (35 USC § 285).

11.6 Inequitable conduct: the duty of candour, good faith and honesty to the US PTO

A US patent can be held unenforceable if a patentee committed inequitable conduct during prosecution of the patent. A holding of inequitable conduct can lead to a finding that a patent infringement action is an exceptional case meriting an award of attorneys' fees to the accused infringer. Inequitable conduct, however, should not be asserted lightly. It is a defence that must be pleaded with particularity (Fed R Civ P 9).

Patent prosecution in the United States is an *ex parte* process. As a result, the patent applicant owes a duty of candour, good faith and honesty to the PTO (37 CFR

---

5   Data from PTO *Ex Parte* Reexamination Filing Data – December 31, 2007.

6   Andrew S Baluch & Stephen B Maebius, "The Surprising Efficacy of *Inter Partes* Reexaminations: An Analysis of the Factors Responsible For Its 73% Patent Kill Rate and How To Properly Defend Against It", Foley & Lardner LLP (2008).

§ 1.56(a)). A patent applicant may breach that duty by failing to disclose information – or presenting incomplete or misleading information – that is material to the examination of the application.

The duty of candour applies broadly to any person substantively involved in the prosecution of the patent application, including the named inventors and prosecuting attorneys (37 CFR § 1.56(c)).

Inequitable conduct is shown by a failure to disclose material information to the US patent examiner coupled with an intent to deceive or mislead. Both materiality and intent must be proved by clear and convincing evidence. "Once thresholds of materiality and intent have been established, the court conducts a balancing test and determines whether the scales tilt to a conclusion that 'inequitable conduct' occurred" (*Critikon, Inc. v Becton Dickinson Vascular Access, Inc.*, 120 F.3d 1253, 1256 (Fed Cir 1997)). In other words, the "more material the information misrepresented or withheld by the applicant, the less evidence of intent will be required in order to find that inequitable conduct has occurred" (*Li Second Family Ltd. P'ship. v Toshiba Corp.*, 231 F.3d 1373, 1378 (Fed Cir 2000)).

Information is material if there is a substantial likelihood that a reasonable examiner would have considered the information important in deciding whether to allow the application to issue as a patent (*Li Second Family*, 231 F.3d at 1379–80 (citation omitted); see also *Digital Control Inc. v Charles Mach. Works*, 437 F.3d 1309, 1314–16 (Fed Cir 2006) (finding that the 'reasonable examiner' standard applies to inequitable conduct determinations)). "Materiality is not limited to prior art but embraces any information that a reasonable examiner would be substantially likely to consider important in deciding whether to allow an application to issue as a patent" (*GFI, Inc. v Franklin Corp.*, 265 F.3d 1268, 1274 (Fed Cir 2001)).

In inequitable conduct cases, the Federal Circuit cautions that intent to deceive need not be, and usually is not, proven by direct evidence. Thus, an intent to deceive may be inferred from the surrounding circumstances or by "a showing of acts, the natural consequences of which are presumably intended by the actor" (*Molins PLC v Textron, Inc.*, 48 F.3d 1172, 1180 (Fed Cir 1995); *Merck & Co. v Danbury Pharmacal, Inc.*, 873 F.2d 1418, 1422)(Fed Cir 1989)).

Intent to deceive may also be inferred from high materiality and evidence that the patentee "should have known" of the materiality (*Critikon*, 120 F.2d at 1257). "[A] trial court may infer deceptive intent based on a showing that a patentee withheld references with which it was intimately familiar and which were inconsistent with its own patentability arguments to the PTO" (*AGFA Corp. v Creo Prods. Inc.*, 451 F.3d 1366, 1378 (Fed Cir 2006)). However, any inference of intent drawn from the evidence must be the "single most reasonable inference". *Star Scientific, Inc. v R. J. Reynolds Tobacco Co.*, 537 F.3d 1357, 1365–66 (Fed Cir 2008); *Scanner Techs. Corp. v ICOS Vision Sys. Corp. N.V.*, 528 F.3d 1365, 1375–79 (Fed Cir 2008). Proof of inequitable conduct as to one claim in a patent renders all claims of that patent unenforceable.

About the authors

George A Ballas

Managing Partner, Ballas, Pelecanos & Associates

george.ballas@balpel.gr

George A Ballas is a senior and managing partner of Ballas, Pelecanos & Associates – the Greek member firm of Counterforce, a network of law firms established by the International Chamber of Commerce's Counterfeit Intelligence Bureau. He is a European patent attorney and a solicitor and barrister before the Athens and Piraeus Court of Appeals, the Supreme Court of Greece and the Council of State. He is a graduate of the University of Athens (LLB) and the University of Paris.

Mr Ballas has more than 33 years' experience in IP law issues, covering both administration of IP rights and litigation of infringements – including counterfeits. As a trademark and patent specialist, he represents and advises a large number of well known multinational companies.

Mr Ballas has served as the general legal counsel of Fiat Auto Hellas SA and is the lead outside counsel for Greece of Microsoft Corporation.

Thomas Bopp

Partner, Gleiss Lutz

thomas.bopp@gleisslutz.com

Thomas Bopp is a partner in the patents department at Gleiss Lutz, a leading international law firm with more than 240 lawyers and offices in Berlin, Frankfurt, Munich, Stuttgart and Brussels. Since joining the Stuttgart office of Gleiss Lutz in 1989, Dr Bopp has accumulated almost 20 years' experience in the field of patent litigation. He has been a partner at the firm since 1992.

Giovanni Francesco Casucci

Managing Partner, Casucci Law Firm

gcasucci@casucci.biz

Giovanni F Casucci is the founder of the Casucci Law Firm, which is a new firm specialising exclusively in IP litigation. It has operated from two offices in Milan and Verona since May 2005, and since January 2009 with a third office in Venice. Giovanni has worked in the IP enforcement sector for almost 15 years, having also been head of the IP department in Clifford Chance Italy for three years. He has particular experience in patent litigation, design protection and border enforcement. He is also Professor of IP at the biotech faculty of the University of Milano Bicocca and member of the permanent faculty of the MBA programmes of the Politecnico of Milano University. Giovanni frequently acts as a trainer in training programmes for IP judges, organised by the EPO Academy and the WIPO Academy.

Vasco Stilwell d'Andrade

IP law consultant, Morais Leitão, Galvão Teles, Soares da Silva & Associados, RL (MLGTS)

vsandrade@mlgts.pt

Vasco Stilwell d'Andrade is a graduate both in Law from the *Universidade Lusíada de Lisboa* and in International Relations and Modern History from the University of St Andrews, Scotland. He also completed an LLM Masters of Law at the University of Edinburgh, where he specialised in international law and intellectual property law. He has been working in the intellectual property field since 2000 and currently works as an IP law consultant with MLTGS.

Paul England
Solicitor, Herbert Smith LLP
paul.england@herbertsmith.com

Before turning to practise law, Paul England was awarded an honours degree in chemistry at Edinburgh University, followed by a doctorate in biochemistry and molecular biology from Oxford University. His doctoral thesis was based on commercially sponsored research involving the genetic engineering of recombinant proteins.

Paul also has direct experience of the life sciences and pharmaceuticals sectors, having been involved with the Human Genome Organisation (HUGO) and working as a chemist with the former Wellcome Foundation, now GSK. Consequently, Paul's principal practice is in life sciences and pharmaceuticals, and other high-tech patent litigation.

Paul has been involved in a number of high-profile patent litigation cases, acting for Roche, AstraZeneca and Daiichi Sankyo, amongst others. He has published widely on patents and other IP matters and is cited by Chambers and Partners 2009. Paul is a member of the International Association for the Protection of Intellectual Property (AIPPI).

Gonçalo da Cunha Ferreira
Partner, Garrigues
goncalo.cunha.ferreira@garrigues.com

Gonçalo da Cunha Ferreira graduated in law at the Universidade Moderna in Lisbon and is a member of the Portuguese Bar Association. He has completed a postgraduate course in Strategy Management (INDEG/ISCTE) and several workshops in Mediation and Arbitration

(ASIPI/INTA/WIPO). Gonçalo has been working in IP matters as a consultant since 1985, first at Gastão da Cunha Ferreira & Associados, Lda, and then continuing with its incorporation into Garrigues in 2006. He is a member, and regular attendee of conferences and seminars, of several national and international associations such as ACPI, AIPPI, ASIPI, ECTA, FICPI, INTA or PTMG. Gonçalo da Cunha Ferreira is now head of the Portuguese IP/IT department of Garrigues and director of the Iberian IP prosecution practice.

James F Haley
Partner, Ropes & Gray LLP
james.haley@ropesgray.com

Ropes & Gray is a leading international law firm with six offices in the United States and two in Asia. Jim Haley has been a partner at Ropes & Gray (formerly Fish & Neave) since 1983. He is active in patent litigation, oppositions, interferences, re-examinations and IP rights management and he specialises in biopharmaceuticals and pharmaceuticals. Jim is formerly co-head of the IP corporate practice group and was part of the trial team that won a jury verdict of more than $100 million for the Massachusetts Eye & Ear Infirmary in 2006.

Rainer Hilli
Partner, Roschier, Attorneys Ltd
rainer.hilli@roschier.com

Roschier has offices in Finland and in Sweden, and an alliance (RoschierRaidla) in the Baltic states. Rainer Hilli is head of Roschier's intellectual property and technology practice and specialises in all types of intellectual property matters. Rainer has worked in the field of intellectual property for more than 20 years. He regularly advises on patent matters, as well as other intellectual property, media and information technology issues including general advice, licensing, regulatory, transactions and litigation. His experience includes multi-jurisdictional patent litigation, especially in the fields of pharmaceuticals, electronics and telecommunications. Rainer is currently president of AIPPI Finland.

Henrik Holzapfel

Associate, Gleiss Lutz

henrik.holzapfel@gleisslutz.com

Dr Henrik Holzapfel is an associate in the patents department at Gleiss Lutz, a leading international law firm with more than 240 lawyers and offices in Berlin, Frankfurt, Munich, Stuttgart and Brussels. Since joining the Stuttgart office of Gleiss Lutz in 2005, Dr Holzapfel has gained significant experience in the field of patent litigation.

José Massaguer

Partner, Uría Menéndez

jmf@uria.com

José Massaguer is a partner in the Madrid office of Uría Menéndez and a chaired professor of commercial law at the Universidad Pompeu Fabra. He focuses his practice on copyright, intellectual property, unfair competition and commercial law. He regularly advises and represents domestic and foreign companies in proceedings relating to patents, trademarks, industrial models and drawings, copyright, know-how and other copyright and intellectual property rights, as well as in matters relating to unfair competition and advertising.

William J McCabe

Partner, Ropes & Gray LLP

william.mccabe@ropesgray.com

Ropes & Gray is a leading international law firm with six offices in the United States and two in Asia. Bill McCabe has been a partner at Ropes & Gray (formerly Fish & Neave) since 1997. He has specialised in IP litigation for 20 years and is currently head of the IP litigation practice group. Bill was part of the trial team in *Symbol Technologies et al. v Lemelson Foundation*, a case

widely viewed as being the biggest 'defence' win in US patent litigation history.

Sebastian Moore

Partner, Herbert Smith LLP

sebastian.moore@herbertsmith.com

Sebastian Moore has a chemistry degree and has worked on a wide range of life science matters. As well as running large multi-jurisdictional pharmaceutical patent actions (including the first streamlined patent action to come to trial), he has advised a number of pharmaceutical and biotechnology companies on licensing and freedom-to-operate issues. He has also conducted trade mark, telecoms and information technology litigation.

Sebastian is a solicitor advocate with rights of audience in all civil proceedings. He is a guest lecturer at the University of London and a tutor on the Oxford University Diploma in Intellectual Property Law and Practice.

Sebastian has also been involved in drafting the submissions on behalf of the Intellectual Property Lawyers Association (IPLA) to the EU Commission on the reform of the European Patent System.

Alexandra Neri

Partner, Herbert Smith LLP

Alexandra.Neri@herbertsmith.com

Alexandra Neri is a partner at Herbert Smith LLP, Paris, where she heads the IP/IT group. She is an expert in contentious and non-contentious IP issues and is both a mediator and an arbitrator for the World Intellectual Property Organisation. Alexandra advises and litigates in all areas of intellectual property, including copyrights and neighbouring rights, trade marks, patents and industrial designs. Her practice also encompasses information technology, including advising on IT projects, international data protection and cross-border data flows, IT security, e-commerce and dematerialisation. Her experience in IT litigation includes domain name litigation and online copyright and trademark infringement. According to the Chambers Global Guide 2008 "Alexandra Neri is an 'extremely good fighter in disputes', whose presence in the IT and IP sectors has grown considerably in recent years."

Masahiro Otsuki
Partner, Abe, Ikubo & Katayama
otsuki@aiklaw.co.jp

Masahiro Otsuki is a partner in the law department at Abe, Ikubo & Katayama, a leading law firm in intellectual property law and corporate related laws in Japan. Mr Otsuki is a lawyer admitted in Japan and New York. He was also an adjunct lecturer at the University of Tokyo School of Law. Mr Otsuki deals with a wide variety of fields, such as finance law (mergers and acquisitions), insolvency law and antitrust law, as well as intellectual property rights law, based on his international working experience at investment banks and law firms in Europe and the United States. In the area of intellectual property, in addition to being engaged in much international litigation and providing legal opinions and drafting agreements, Mr Otsuki provides consultations on patents, trade marks and other intellectual property rights to Japanese and overseas companies.

Loukia K Papas
Associate, Ballas, Pelecanos & Associates
Loukia.papas@balpel.gr

Loukia K Papas is an associate with Ballas, Pelecanos & Associates in Athens. She is a graduate of Thessaloniki University Law School and has extensive experience in litigation of industrial property rights, as well as in competition, commercial, contract and sports law issues. She has worked for the Industrial Property Organization (OBI) for several years, has advised many multinational corporations on patent litigation cases and published several articles and contributions in legal books in the field of patent law enforcement.

Ingrid Pi
Senior Associate, Uría Menéndez
IPA@uria.com

Ingrid Pi is a senior associate practising in all areas of intellectual property law. Her practice extends to litigation and non-litigation issues in connection with the defence of patents, plant varieties, trade marks, designs and other intellectual property issues as well as unfair competition. She has extensive experience in patent law. For the past few years she has been involved in a number of patent infringement and nullity proceedings relating to patents in all technical fields, including proceedings to obtain and to preserve evidence and interim injunction proceedings. In addition, she has extensive knowledge of regulatory issues relating to medicinal products. She also lectures on intellectual property in postgraduate programmes and specialised seminars.

Peter-Ulrik Plesner
Partner, Plesner
pup@plesner.com

Peter-Ulrik Plesner is a partner in the IP department at Plesner, which is one of the leading Danish law firms. Peter-Ulrik works on IP matters and especially patent litigation. He has experience in litigation in all Danish courts, including the Supreme Court. He is also Chairman of the Danish AIPPI Group and, as such, is a member of the AIPPI Executive Committee. Peter-Ulrik has presented several AIPPI reports and lectures to meetings in the Nordic AIPPI group. He has also lectured in several AIPPI-related matters, especially concerning patents. He represents a number of national and international manufacturers of pharmaceutical products.

Sture Rygaard

Partner, Plesner

sry@plesner.com

Sture Rygaard is partner in the IP department at one of the leading Danish law firms, Plesner. Sture's primary areas of practice are patents, trade marks, designs, life science law and marketing law, as well as distribution and technology transfer agreements. Litigation and interlocutory injunctions proceedings in these fields of law constitute a large part of his work.

Sture is Chairman of the Board of the Danish Anti-Counterfeiting Group. He is Master of Law (LLM) from Cambridge University, England. He is a member of the Danish Association for the Protection of Industrial Rights and AIPPI and has given lectures to meetings in the Nordic AIPPI group.

Sascha D Salomonowitz

Junior Partner, Schoenherr

s.salomonowitz@schoenherr.at

Sascha Daniel Salomonowitz is junior partner at Schoenherr, a leading corporate law firm in Austria and Central Eastern Europe with more than 300 fee earners and offices in 12 countries. Schoenherr is considered to be among the leading law firms in the field of intellectual property and represents clients in contentious and non-contentious patent, trade mark, design, copyright and unfair competition matters as well as anti-piracy/anti-counterfeiting issues.

Sascha represents clients in patent litigation as well as technology-driven transactions and commercialisation of intellectual property through licensing, joint R&D, and distribution agreements. He is particularly active for international and European innovative market leaders, and represents various companies in the field of electronics, consumer goods, chemistry and pharmaceuticals. Sascha was involved in some of the first house-search orders granted in Austria in a civil IP infringement case based on the EU Enforcement Directive.

Paweł Siekierzyński

Partner, Gleiss Lutz

Paweł Siekierzyński@gleisslutz.com

Paweł Siekierzyński is a partner in the corporate department at Gleiss Lutz, a leading international law firm with seven offices in four countries. Mr Siekierzyński has particular experience of cross-border mergers and acquisitions transactions. He is also head of the IP/IT team in Gleiss Lutz's office in Warsaw.

Caroline Thufason

Assistant attorney, Plesner

ctn@plesner.com

Caroline Thufason is assistant attorney in the IP department at Plesner, one of the leading Danish law firms. Caroline works on IP matters, especially patent litigation, and has assisted both Peter-Ulrik Plesner and Sture Rygaard in the preparation of patent litigation since 2006. She is Master of Law from the University of Copenhagen and has studied intellectual property law at the London School of Economics. She is a member of the Danish Association for the Protection of Industrial Rights and AIPPI.

Ignace Vernimme

Partner, Stibbe

Ignace.vernimme@stibbe.com

Ignace Vernimme has special expertise in IP law and related areas within the pharmaceutical, metallurgic, comic strip, fashion and food industries. His activities encompass in particular patents, technology transfer, trade marks and copyright, designs, know-how protection and the specific rules applicable to pharmaceutical products. He was admitted to the Brussels Bar in 1987 and joined Stibbe as a partner in 2007. He is the author of many publications in the field of IP law.

John Whelan

Partner, A&L Goodbody

jwhelan@algoodbody.ie

John Whelan is head of intellectual property and technology at A&L Goodbody. His practice areas include both the contentious and non-contentious aspects of intellectual property, information technology, telecommunications and life sciences.

Before joining A&L Goodbody as a partner in 2003, John previously worked with a leading international law firm in London and Hong Kong. His return to Ireland coincided with the establishment of the Irish Commercial Court in January 2004 and he has been involved in most of the significant IP cases before that court since its inception, including a number of complex and high-profile patent proceedings.

John has spoken at conferences on IP topics in a number of jurisdictions and is recommended as a lawyer of choice for IP and IT matters by a number of international legal publications.

David Wilson

Partner, Herbert Smith LLP

david.wilson@herbertsmith.com

David Wilson joined Herbert Smith as a partner in 2007 from a leading city IP firm where he had been a partner since 2001.

David's practice has focused on complex patent litigation involving multi-jurisdictional disputes and work involving technical subject matter in the life sciences sector and chemical industries. In addition, he advises clients in relation to trademarks, regulatory and non-contentious (contractual, licensing and corporate) matters.

David has a first-class honours degree in chemistry and, prior to pursuing a career as a lawyer, he worked in the chemical and life sciences (immunodiagnostics) industry. David also qualified as a barrister before joining his previous firm in 1994. He is recognised by the major legal directories as a leader in the intellectual property and, in particular, patent litigation fields.